621·38   7-95

D1148679

Transmission Principles Technicians

D C Green
M Tech, CEng, MIERE
Formerly Senior Lecturer in Telecommunication

BANFF AND BUCHAN COLLEGE
B0006341

Longman Group UK Limited,
Longman House, Burnt Mill, Harlow,
Essex CM20 2JE, England
and Associated Companies throughout the world.

© Longman Group UK Ltd 1988

All rights reserved; no part of this publication
may be reproduced, stored in a retrieval system,
or transmitted in any form or by any means, electronic,
mechanical, photocopying, recording, or otherwise,
without either the prior written permission of the
publishers or a licence permitting restricted copying
issued by the Copyright Licensing Agency Ltd, 33—34
Alfred Place, London, WC1E 7DP.

First published 1988

British Library Cataloguing in Publication Data

Green, D.C.
  Transmission principles for technicians.
  1. Telecommunication systems
  I. Title
  621.38   TK5101

ISBN 0-582-99461-6

6,015
621.38

Produced by Longman Group (FE) Limited
Printed in Hong Kong

C.9.88

Preface

This book provides an introduction to the basic principles and practice of telecommunication line systems. The intended readership in the United Kingdom is the student of either the Business and Technician Education Council (BTEC) or the City and Guilds of London Institute units in Transmission Principles at both levels II and III.

A study of the book should give the reader the understanding of the operation of both analogue and digital transmission systems that is necessary for all telecommunication technicians, whether their particular interests lie in the field of data, line or radio communications, or in that of switching systems.

The earlier chapters of the book present, using a minimum of mathematics, the basic concepts of signals, modulation, noise, and transmission lines. These concepts are then employed, in the later chapters, to introduce the reader to practical transmission systems.

Many worked examples are provided in the text to illustrate the principles which have been discussed. At the end of the book will be found a number of short answer questions. The intention is that the reader will use these problems to check his understanding of each topic.

D.C.G.

Contents

1 Signals

In telecommunication engineering, media are provided for the transmission of intelligence from one point to another; for example, telephone conversations pass through telephone cables and radio programmes are broadcast through the atmosphere. It is necessary to ensure that sufficient information is available at the receiving end of a system to allow the intelligence to be understood and appreciated by the person receiving it. The requirements demanded of the transmission media depend upon the type of intelligence to be transmitted, but for the transmission of speech, music and television the main requirement is that sufficient frequencies are retained in the transmitted waveform to permit the received sound and picture to be intelligible and, in the case of music and television, to be enjoyable also. It is therefore necessary to have an appreciation of the range of frequencies produced by the human voice, by musical instruments and by television systems, and the frequency range over which the human ear is capable of responding.

The Voice and Speech

A current of air expelled by the lungs passes through a narrow slit between the vocal cords in the larynx and causes them to vibrate. This vibration is then communicated to the air via various cavities in the mouth, throat and nose. The shape and size of the nose cavities are more or less fixed, but the mouth and throat cavities can have their shapes and dimensions considerably changed by the action of the tongue, lips, teeth and the throat muscles. The frequency at which a particular cavity allows the air to vibrate most freely depends upon the shape and dimensions of the cavity and these can readily be adjusted by the movement of the lips, tongue and teeth. The pitch of the spoken sound depends upon both the length and tension of the vocal cords and the width of the slit between them. The length of the vocal cords varies from person to person; for example a woman will have shorter vocal cords than a man, while the tension and distance apart of the vocal cords are under the control of muscles.

When a person speaks, his vocal cords vibrate and the resulting sounds, which are rich in harmonics, but of almost constant pitch, are carried to the cavities in the mouth, throat and nose. Here the sounds are given some of the characteristics of the desired speech by the emphasizing of some of the harmonics contained in the sound waveforms and the suppression of others. Sounds produced in this way are the vowels, a, e, i, o and u, and contain a relatively large amount of sound energy. Consonants are made with the lips, tongue and teeth and contain much smaller amounts of energy and often include some relatively high frequencies.

The sounds produced in speech contain frequencies which lie within the frequency band 100—10 000 Hz. The pitch of the voice is determined by the fundamental frequency of the vocal cords and is about 200—1000 Hz for women and about 100—500 Hz for men. The tonal quality and the individuality of a person's voice are determined by the higher frequencies that are produced.

The power content of speech is small, a good average being of the order of 10—20 microwatt. However, this power is not evenly distributed over the speech frequency range, most of the power being contained at frequencies in the region of 500 Hz for men and 800 Hz for women.

Music

The notes produced by musical instruments occupy a much larger frequency band than that occupied by speech. Some instruments, such as the organ and the drum, have a fundamental frequency of 50 Hz or less while many other instruments, for example the violin and the clarinet, can produce notes having a harmonic content in excess of 15 000 Hz. The power content of music can be quite large. A sizeable orchestra may generate a peak power somewhere in the region of 90—100 watts while a bass drum well thumped may produce a peak power of about 24 watts.

Hearing

When sound waves are incident upon the ear they cause the ear drum to vibrate. Coupled to the ear drum are three small bones which transfer the vibration to a fluid contained within a part of the inner ear known as the cochlea. Inside the cochlea are a number of hair cells and the nerve fibres of these are activated by vibration of the fluid. Activation of these nerve fibres causes them to send signals, in the form of minute electric currents, to the brain where they are interpreted as sound.

The ear can only hear sounds whose intensity lies within certain limits; if a sound is too quiet it is not heard and, conversely, if a sound is too loud it is felt rather than heard and causes discomfort or even pain. The minimum sound intensity, measured in pascals (1 Pa = 1 N/m^2), that can be detected by the ear is known as the 'threshold of hearing or audibility' and the sound intensity that just produces a feeling of discomfort is known as the 'threshold of feeling'. The ear is not, however, equally sensitive at all frequencies, as shown in Fig. 1.1. In this diagram curves have been plotted showing how the thresholds of audibility and feeling vary with frequency for an average person.

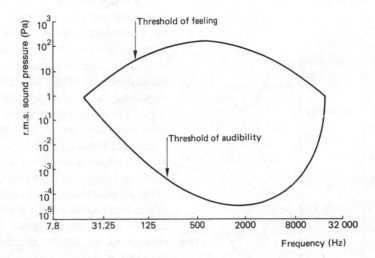

Fig. 1.1 The thresholds of audibility and feeling

It can be seen that the frequency range over which the average human ear is capable of responding is approximately 30−16 500 Hz, but this range varies considerably with the individual. The ear is most sensitive in the region of 1000 to 2000 Hz and becomes rapidly less sensitive as the upper and lower limits of audibility are approached. The limits of audibility are clearly determined not only by the frequency of the sound but also by its intensity. See, for example, the increase in the audible frequency range when the sound intensity is increased from, say, 1×10^{-3} Pa to 1×10^{-2} Pa. At the upper and lower limits of audibility the thresholds of audibility and feeling coincide and it becomes difficult for an observer to distinguish between hearing and feeling a sound.

The Transmission of Sound over a Simple Telephone Circuit

The intensity of a sound wave rapidly diminishes as it travels away from the source producing it, and if conversation over a long distance is desired a telephone circuit becomes necessary.

The arrangement of a simple, *unidirectional* telephone circuit is shown in Fig. 1.2.

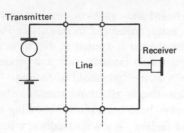

Fig. 1.2 A simple unidirectional speech circuit

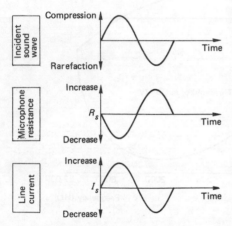

Fig. 1.3 The relationship between the sound incident on a microphone, the microphone resistance, and the current flowing to line. R_s = microphone resistance when no sound is incident upon it. I_s = steady current to line when no sound is incident on the microphone

It is sufficient to consider the microphone as a device whose electrical resistance varies in accordance with the waveform of the sound incident upon it, and to consider the receiver as a device which vibrates when a varying current is passed through it and in so doing produces sound waves having the same waveform as the current.

When no sound is incident upon the microphone, the resistance of the microphone is constant and a steady current flows into the line from the battery. When a person speaks into the microphone, its resistance varies in the same way as the speech waveform and so does the current flowing to line. For example, during one half-cycle of the speech waveform the resistance of the microphone is decreased and so the line current is increased, while during the next half-cycle the microphone resistance is increased and the line current decreased. Thus the line current is continuously varying about its steady value (Fig. 1.3). The varying line current passes through the receiver at the distant end of the line and causes the receiver to vibrate and reproduce the original speech.

For a *two-way* conversation this simple arrangement would have to be duplicated, the second circuit having the positions of the microphone and battery and the receiver reversed. Such an arrangement would be uneconomic since it would require two pairs in a telephone cable, and telephone cables are very expensive. A further disadvantage of the simple circuit is that as the length of line is increased, the variation in the resistance of the microphone becomes an increasingly small fraction of the line resistance. This means that the magnitude of the changes in line current decreases with increase in line length until the current changes can no longer operate the receiver.

A telephone circuit which overcomes these disadvantages is shown in Fig. 1.4. In this circuit, each microphone is connected to the line via a transformer but the two receivers are directly connected to line. When a person speaks into either of the two microphones, the resulting changes in microphone resistance cause a relatively large varying current to flow in the local microphone circuit. In passing through the transformer primary winding, this varying current induces an e.m.f. in the secondary winding and the induced e.m.f. drives a varying current, having the same waveform, to line. At the distant end of the line the varying current flows in the receiver and causes it to vibrate and so produce sound waves which are similar to the original speech.

The circuit given in Fig. 1.4 has the disadvantage of requiring a separate battery at each telephone. The vast majority of telephone

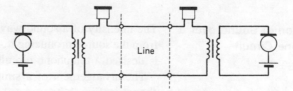

Fig. 1.4 A simple two-way speech circuit

circuits are connected to a local telephone exchange and these circuits are all operated from a *central battery*. The basic principle of the central battery system is illustrated by Fig. 1.5. A large capacity secondary cell battery is installed at the local telephone exchange which supplies current to the lines when the telephone handsets are lifted from their rests. When either of the microphones is spoken into, a speech-frequency current is superimposed upon the battery current and is transmitted, via the telephone exchange, to the other telephone.

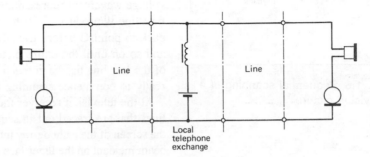

Fig. 1.5 The central battery system

The inductor is connected in series with the exchange battery to prevent the speech-frequency currents entering the battery. In practice, a more complex telephone circuit is needed because the circuit as shown would suffer from the speech-frequency currents generated by the microphone flowing in the associated receiver. The sound picked up by the microphone would be clearly audible in the receiver to give excessive sidetone.

The telephone network of Great Britain is divided into local lines, junctions and trunks. *Local lines* connect the individual telephone subscribers to their local telephone exchange; *junctions* are short circuits that connect nearby telephone exchanges; and *trunks* connect more distant exchanges.

Long-distance telephone lines are extremely expensive and it is not economically possible to connect every exchange in the network to every other exchange; direct trunks are only provided between two exchanges when justified by the traffic carried. The vast majority of trunk circuits are routed over one or more multi-channel telephony systems. International circuits may be routed over submarine cable, microwave radio or satellite multi-channel systems, or sometimes over high-frequency (3–30 MHz) radio links.

Television Signals

A television picture is divided vertically into 625 lines, each of which is effectively subdivided into a number of sections. A television picture is therefore divided into a number of elemental areas. If the elemental areas are small enough each will have a constant brightness; for monochrome television, the brightest areas are white, the darkest

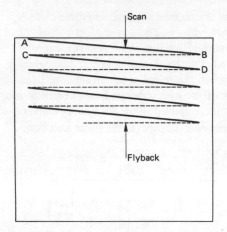

Fig. 1.6 Sequential scanning of a television picture

black, and all other areas are various shades of grey. The action of the television camera is to convert the brightness of each area into a voltage whose amplitude is proportional to the brightness.

The camera is unable to transmit all the voltages simultaneously, and so the system is arranged to transmit them sequentially. The picture is *scanned* by the camera in a series of lines as in Fig. 1.6. The scan starts at the top left-hand corner of the picture, point A, and travels along the first line to point B. As each elemental area is scanned a proportional instantaneous voltage is transmitted and a voltage waveform representing the variation of brightness along the first line is produced. At the end of this line the scan flies rapidly back to point C before travelling along the second line to point D, and so on until the entire picture has been scanned. When the end of the last line has been reached the scan is moved back to point A ready to commence scanning the next field.

At the television receiver the picture is built up from a number of lines that are traced out in sequence by a spot of light travelling on the screen of the cathode-ray tube. The spot is produced by an electron beam incident on the inner face of the screen, and to obtain the various shades of grey, and black and white demanded by a particular picture the brightness of the spot is modulated as it travels by the picture signal. For the picture to be reproduced correctly it is essential for the two scans (camera and receiver) to be in synchronism. The picture signal produced by the television camera is therefore accompanied by *synchronizing pulses* (Fig. 1.7). The persistence of vision of the human eye enables the viewer to see a complete picture and not a moving spot of light. The fields must be repeated at a frequency high enough for the slight changes in successive fields to give the impression of movement without apparent flicker. The necessary field frequency is reduced by 50% since *interlaced scanning* is used; here the odd lines, 1, 3, 5, 7, etc., are scanned first and the even lines afterwards.

The bandwidth required for a television picture signal depends upon a number of factors, such as the number of lines that make up the picture, the number of fields transmitted per second and the durations

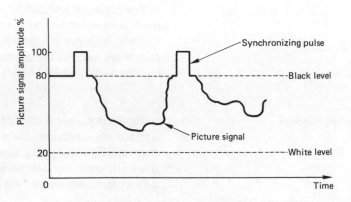

Fig. 1.7 Television picture waveform

of the synchronizing pulses. In practice, the normal bandwidth provided is 5.5 MHz.

A colour television picture signal consists of a brightness (*luminance*) component, which corresponds to the monochrome signal previously described, plus colour (*chrominance*) information which is transmitted as the amplitude-modulation sidebands of two colour *subcarriers* which are of the same frequency (approximately 4.434 MHz) but are 90° apart. No extra bandwidth is needed to accommodate the colour information.

Telegraph Signals

Telegraphy is the passing of messages by means of a signalling code such as the *Morse code* and the *Murray code*.

In the *Morse code* characters are represented by a combination of *dot* signals and *dash* signals; the difference between a dash and a dot is one of time duration only, a dash having a period three times that of a dot. Spacings between elementary signals, between letters and between words are also distinguished from one another by different time durations. As an example, Fig. 1.8 shows the word BAT in Morse code. The Morse code is not convenient for use with automatic-printing receiving equipment because the number of signal elements needed to indicate a character is not the same for all characters, and the signal elements themselves are of different lengths.

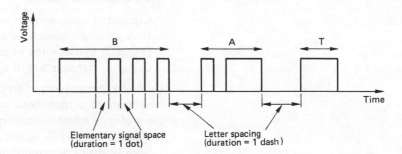

Fig. 1.8 BAT in Morse code

In the *Murray code*, or *International Alphabet 2* (IA2), all characters have exactly the same number of signal elements and the signal elements are of constant length. Each character is represented by a combination of five signal elements that may be either a *mark* or a *space*. A mark is represented by a negative potential or the presence of a tone, and a space is represented by a positive potential or the absence of a tone. Figure 1.9 shows the letters B, T, R and Y in IA2 code. The IA2 code is used for teleprinter systems.

Telegraph speed is measured in terms of a unit known as the *baud*. The baud speed of a telegraph signal is the reciprocal of the time duration of the shortest signal element employed. In the Morse code the dot is the shortest signal element; in the IA2 code the elements are all of the same length. The bandwidth required for the transmission

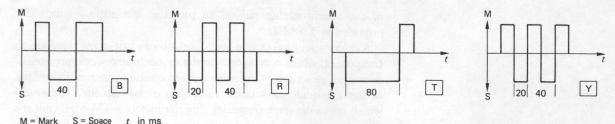

M = Mark   S = Space   *t* in ms

Fig. 1.9 B, T, R and Y in Murray code

of a signal in Morse code depends upon the number of words transmitted per minute and has not been standardized, but generally lies in the range 100–1000 Hz. Teleprinters are normally operated at a speed of 50 bauds, and this means that the time duration of a mark, or of a space, is 1/50 second, or 20 ms. The bandwidth required to transmit a teleprinter signal depends upon the characters sent, but the maximum bandwidth is demanded when alternate marks and spaces are transmitted, i.e. letters R and Y. The periodic time of one cycle of the waveforms for R and Y is 40 ms, and so the fundamental frequency of the waveform is 1000/40, or 25 Hz.

Facsimile Telegraphy

Facsimile telegraphy or *fax* is the transmission of photographs, diagrams, documents, etc. Facsimile terminals are classified by the CCITT into four groups.

Group 1: transmission of an A4 document in 6 minutes.
Group 2: transmission of an A4 document in 3 minutes.
Group 3: transmission of an A4 document in under 1 minute.
Group 4: transmission of a document over a digital network.

Currently, the majority of terminals in use are group 3 types. All fax terminals, other than those in group 4, are designed to work over the analogue telephone network but their operation may be either analogue or digital in nature.

With an analogue terminal the document to be transmitted is placed into the machine and then it is continuously scanned in a number of lines. The scanning process produces a continuous signal that is proportional to the lightness of the part of the page being scanned at that moment. The maximum signal voltage is generated by a white area being scanned and the minimum voltage by a black area. At the receiving end of the system the received analogue signal is used to control a mechanism that prints a replica of the original document.

In a digital terminal the document is again scanned in a number of lines but each line is now divided into a number of picture elements or *pels*. The shade of grey of each pel is quantized (see p. 134) and then represented by means of a binary coded signal. Basic group 3 terminals transmit their data at either 4800 or 2400 bits per second using a form of digital modulation (see p. 92) but some terminals may well be able to transmit at even higher bit rates.

Data Signals

Nowadays, digital computers are widely used by many organizations for scientific and engineering computations, and for commercial data processing applications, such as the calculation and addressing of bills and the compilation of company statistics and records. Much of the data to be processed and perhaps stored by a computer is originated at, and the results are required by, offices and laboratories that are not located at the same geographical point as the computer installation. This means that there is an ever-increasing demand for data links that will enable branch offices to communicate with a central computer. Other examples of data links are electronic mail, the checking of credit cards, and bank cash-card dispenser points. The basic arrangement of a data system is shown by Fig. 1.10. Communication with the computer is carried out via a keyboard, magnetic tape, or a floppy or hard disc since a computer needs its input data in binary form.

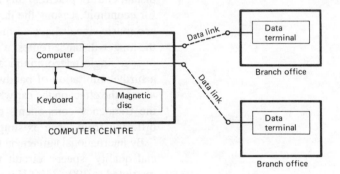

Fig. 1.10 Communication with a computer

Access to the computer will be possible from several different points within the computer centre and also from a number of branch offices linked by telephone line to the centre. Sometimes more than one computer may be integrated within the data network of a particular organization and in such cases there will also be a need for two computers to be able to communicate directly with one another.

Generally, a data system requires more characters than a telegraphy system and this means that the IA2 code is usually inadequate. The vast majority of data terminals employ the *International Alphabet 5* code (IA5); this is a 7-bit code that is commonly known as the ASCII code (short for American Standard Code for Information Interchange). IBM terminals use the EBDIC (Extended Binary Coded Decimal Interchange) 8-bit code.

The data fed into a computer is in binary form with the binary number 0 being represented by a positive 6 V voltage and binary 1 by −6 V (see Fig. 1.11). Each binary digit, 0 or 1, is known as a *bit*. The bit rate is the rate at which information can be transmitted and this is the number of bits transmitted per second. The data waveform of Fig. 1.11 consists of a d.c. component (equal to the average

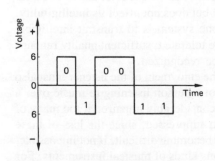

Fig. 1.11 Data waveform

value of the waveform), a fundamental frequency, and a number of harmonics of the fundamental frequency.

The fundamental frequency has its maximum value when the signal consists of alternate 1s and 0s; this maximum value is equal to one-half of the bit rate.

The bit rate of a data signal will be numerically equal to the baud speed whenever the bits are individually transmitted to line. This, however, is only the case for bit rates up to 1200 per second. At higher bit rates bits are transmitted in groups of either two (dibits) or three (tribits) and then the baud speed on the line will be less than the bit rate.

Range of Frequencies in Communication

It has been assumed so far that all the frequencies present in a speech or music waveform are converted into electrical signals, transmitted over the communication system, and then reproduced as sound at the distant end. In practice, this is rarely the case for two reasons. First, for economic reasons the devices used in circuits that carry speech and music signals have a limited bandwidth; second, particularly for the longer-distance routes, a number of circuits are often transmitted over a single telecommunication system and this practice provides a further limitation of bandwidth. It is thus desirable to have some idea of the effect on the ear when it is responding to a sound waveform, the frequency components of which have amplitude relationships differing from those existing in the original sound.

By international agreement the audio frequency band for a 'commercial quality' speech circuit routed over a multi-channel system is restricted to 300–3400 Hz; while for a circuit operated over a high-frequency radio link the bandwidth is only 250–3000 Hz. This means that both the lower and upper frequencies contained in the average speech waveform are not transmitted. To the ear, the pitch of a complex, repetitive sound waveform is the pitch corresponding to the frequency difference between the harmonics contained in the waveform, i.e. the pitch is that of the fundamental frequency. Hence, even though the fundamental frequency itself may have been suppressed, the pitch of the sound heard by the listener is the same as the pitch of the original sound. However, much of the power contained in the original sound is lost. Suppression of all frequencies above 3400 Hz reduces the quality of the sound but does not affect its intelligibility. Since the function of a telephone system is to transmit intelligible speech, the loss of quality can be tolerated; sufficient quality remains to allow a speaker's voice to be recognized.

In the transmission of music the enjoyment of the listener must also be considered. The enjoyment of a person listening to music over a communication system may be considerably impaired if too many of the higher harmonics have been suppressed, since the loss of these harmonics could well result in it becoming difficult, if not impossible, to distinguish between the various kinds of musical instruments. For music circuits routed over line communication systems, a wider bandwidth must be allowed than is allocated to commercial speech circuits. A typical bandwidth in practice is 30 to 10 000 Hz, which makes excellent quality reproduction possible. Land-line music circuits of this kind are used to connect together two BBC studios or a studio and a transmitting station. When music is broadcast in the long and medium wavebands a bandwidth as wide as 30−10 000 Hz cannot be achieved because the wavebands are shared by so many different broadcasting stations.

By international agreement medium waveband broadcasting stations in Europe are spaced approximately 9000 Hz apart in the frequency spectrum, and this means that to make it possible for any particular station to be selected by a radio receiver, without undue interference from adjacent (in frequency) stations, the output sound bandwidth of the receiver cannot be much greater than 4500 Hz. Thus the effective bandwidth of a medium wave broadcast transmission, be it music or speech, is of the order of 50−4500 Hz. Sound broadcast stations in the high-frequency band are allowed an r.f. bandwidth of 10 kHz. This provides an audio bandwidth of 50−5000 Hz. For various reasons the same selectivity is not demanded of very high frequency (vhf) sound broadcast receivers or of ultra high frequency (uhf) television receivers (625 line), and consequently the bandwidths handled by these receivers are somewhat greater.

Frequency-modulated sound broadcast transmissions in the 88−108 MHz band provide audio signals up to 15 kHz which allows reasonably good quality reception of musical programmes with a receiver of adequate performance.

The video bandwidth allocated to a uhf television signal is limited to 5.5 MHz. This practice leads to some loss of the horizontal definition of the received picture. The bandwidth of the audio signal is 20 kHz.

2 Frequency, Wavelength, and Velocity

Phase Velocity, Frequency and Wavelength

The propagation of a voltage or current wave along a telephone line, or of a radio wave through the atmosphere, is not an instantaneous process but occupies a definite interval of time. There is thus a time lag between the application of a voltage at one end of a line and the detection of a voltage change at a distant point along the line. This means that there will be a *phase difference* between the a.c. voltages existing at two points along a line at a given instant, since the instantaneous voltage at each point is continuously changing. For example, suppose that for a given line the voltage at a point x miles from the sending-end lags behind the sending-end voltage by 90°. Then at a distance of $2x$ miles from the sending-end the voltage will be lagging by $2 \times 90°$ or 180° on the sending-end voltage. The voltage $4x$ miles from the sending-end will be in phase with the sending-end voltage, i.e. a complete cycle of instantaneous values has been completed. The distance $4x$ is known as *one wavelength*, symbol λ (see Fig. 2.1). The wavelength λ is the distance, in metres, between two similar points, in the propagating waveform, e.g. the distance between two successive positive peaks as shown.

Suppose a sinusoidal voltage of frequency f is applied to the input terminals of a transmission line. The periodic time of this waveform is $T = 1/f$ seconds. In a time T seconds the voltage wave will travel along the line a distance equal to T times the velocity v with which it is propagated. During this time the voltage at the sending-end of the line will have described a complete cycle of variation with time. While this happens the wave travels a distance that is the wavelength λ of the signal. Therefore

$$\lambda = vT = \frac{v}{f}$$

or

$$v = \lambda f \tag{2.1}$$

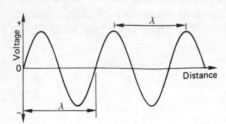

Fig. 2.1 Wavelength of a sinusoidal signal

A radio wave travelling through the atmosphere has its velocity of propagation equal to the velocity of light c. In this book c will be taken as 3×10^8 m/s. The velocity of propagation of a voltage, or current, wave along a transmission line is less than 3×10^8 m/s, the amount of reduction depending upon the inductance and capacitance of the line.

EXAMPLE 2.1

A 30 MHz radio wave is propagated through the atmosphere. Calculate its wavelength.

Solution
From equation (2.1)

$$\lambda = \frac{3 \times 10^8}{30 \times 10^6} = 10 \text{ m} \qquad (Ans.)$$

EXAMPLE 2.2

The BBC broadcasting station Radio 2 is broadcast on a wavelength of 433 m. What is its frequency?

Solution

$$f = \frac{3 \times 10^8}{433} = 693 \text{ kHz} \qquad (Ans.)$$

EXAMPLE 2.3

A 5 MHz radio transmitter is connected to an aerial by a length of transmission line. If the wavelength of a signal on the transmission line is 0.8 times the wavelength of a signal in the atmosphere, calculate the velocity with which a signal is propagated along the cable.

Solution
For the wave in the atmosphere,

$$\lambda_{air} = \frac{c}{f} = \frac{3 \times 10^8}{5 \times 10^6} = 60 \text{ m}$$

Hence, the wavelength in the cable is

$$\lambda_{cab} = 0.8 \times 60 = 48 \text{ m}$$

and therefore the velocity of propagation in the transmission line is

$$v_p = \lambda_{cab} f = 48 \times 5 \times 10^6 = 2.40 \times 10^8 \text{ m/s} \qquad (Ans.)$$

EXAMPLE 2.4

If the maximum frequency tolerance for fixed service stations operating in the frequency band 4.0 to 27.5 MHz is ± 30 parts in 10^6 ($\pm 0.003\%$) and for radio stations operating in the frequency band 40.0 to 70.0 MHz is ± 10 parts in 10^6 ($\pm 0.001\%$), to what frequency variations in hertz do these tolerances correspond for

(a) a fixed service station on 12.3 metres,

(b) a station on 5.5 metres?

At what wavelength does a fixed-service station, having a variation of ±300 Hz, become outside the permitted tolerance?

Solution

$f = v/\lambda$, where f = frequency in hertz, v = velocity in m/s, and λ = wavelength in metres.

(a) $\lambda = 12.3$ m:

$$f = \frac{3 \times 10^8}{12.3} = 24.4 \text{ MHz}$$

The maximum frequency tolerance in this band is ±30 parts in 10^6.

$$\text{Frequency tolerance} = \pm\frac{30}{10^6} \times 24.4 \times 10^6 = \pm732 \text{ Hz} \quad (Ans.)$$

(b) $\lambda = 5.5$ m:

$$f = \frac{3 \times 10^8}{5.5} = 54.55 \text{ MHz}$$

The maximum frequency tolerance in this band is ±10 parts in 10^6.

$$\text{Frequency tolerance} = \pm\frac{10}{10^6} \times 54.55 \times 10^6 = 545.5 \text{ Hz} \quad (Ans.)$$

Last part:

Frequency variation allowed = Frequency × Frequency tolerance

Therefore

$$\pm300 = \text{Frequency} \times \pm\frac{30}{10^6}$$

$$\text{Frequency} = \frac{300 \times 10^6}{30} = 10 \text{ MHz}$$

Therefore

$$\lambda = \frac{3 \times 10^8}{10 \times 10^6} = 30 \text{ m} \quad (Ans.)$$

3 Modulation

The bandwidth required for the transmission of commercial-quality speech is 300—3400 Hz and as long as a physical pair of wires, or two pairs of wires, are provided to connect the two parties to a conversation no problem exists. Telephone cable is, however, very expensive and, together with the associated duct work and manholes, comprises the major part of the cost of linking two points. It is desirable, therefore, to be able to transmit more than one conversation over a given link and thus economize in telephone cable. If a number of telephone conversations were merely transmitted into one end of a line it would not be possible to separate them at the distant end of the linc since each conversation would be occupying the same part of the frequency spectrum. How then can many different conversations be transmitted over a single circuit and yet be separable at the distant end? Two main methods exist: in one, known as frequency-division multiplexing, each conversation is shifted, or translated, to a different part of the frequency spectrum. In the other, known as time-division multiplexing, the conversations each occupy the same frequency band but are applied in time sequence to the common line. Both types of multiplexing improve the efficiency with which line plant is used but tdm is less affected by noise on the line.

Frequency-division Multiplex

To illustrate the principle of a frequency-division multiplex (fdm) system, consider the simple case where it is required to transmit three telephone channels, of bandwidth 300 to 3400 Hz, over a common line. The first of these channels can be transmitted directly over the common line and it will then occupy the band 300 to 3400 Hz. The second and third channels cannot also be transmitted directly over the line because they would be inseparable from the first channel and from each other. Suppose, therefore, that instead these two channels are each passed into a circuit which frequency translates, or shifts, them to the frequency bands 4300 to 7400 Hz and 8300 to 11 400 Hz respectively, before transmission to line. The three channels can now all be transmitted over the common line but since there is a frequency gap of 900 Hz between them no inter-channel interference will occur. At the receiving end of the line, filters separate the three channels and further circuits restore the second and third channels to their original frequency bands. The bandwidth provided by the common circuit must be 300–11 400 Hz.

The frequency translation of a channel to a position higher in the frequency spectrum is known as *modulation* and the circuit which achieves it as a modulator. The particular part of the frequency spectrum to which the channel is shifted is determined by the frequency of the sinusoidal *carrier wave* which is modulated. Modulation may be defined as the process by which one of the characteristics of a carrier wave is modified in accordance with the characteristics of a modulating signal. The restoration of a channel to its original position in the frequency spectrum is known as demodulation, the device is known as a demodulator.

A block schematic diagram of the equipment required for one direction of transmission in the three-channel fdm system just described is shown in Fig. 3.1. It should be noted that since the equipment is unidirectional, it needs to be duplicated for transmission in both directions. Furthermore, the line may be a telephone cable or may be a vhf, uhf or microwave radio link or a communication satellite system.

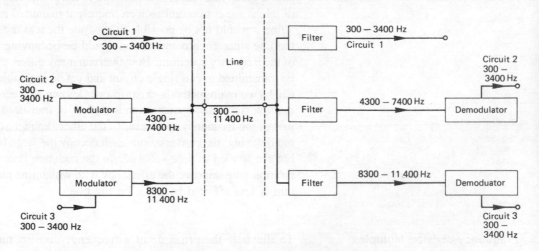

Fig. 3.1 A simple 3-channel fdm system

Frequency translation is also employed for radio and television broadcasting. It is well known that such broadcasts are transmitted by the broadcasting authority and received in the home by means of aerials, but no type of aerial is able to operate at audio frequencies. It is therefore necessary to shift each programme originally produced in the audio-frequency band to a point higher in the frequency spectrum where aerials can operate with reasonable efficiency. Since there is usually a large number of broadcasting stations within a given geographical area, it is necessary to arrange that each station, as far as possible, occupies a different part of the usable frequency spectrum. Hence the programmes radiated by different broadcasting stations are frequency translated to their own, internationally agreed, fixed frequency bands. For example, consider the BBC Radios 1, 2 and 3. Radio 1 is broadcast on frequencies of 1053 and 1089 kHz, Radio 2 on 909 kHz, and Radio 3 on 1215 kHz.

The term *baseband* is often employed to describe the band of frequencies that must be transmited by a fdm system for an adequate signal to appear at the receiving terminal. The baseband signal will usually not contain all the frequencies generated by the source; for example, the speech baseband signal is 300−3400 Hz.

Time-division Multiplex

With time-division multiplex (tdm) a number of different channels can be transmitted over a common circuit by allocating the common circuit to each channel in turn for a given period of time, i.e. at any particular instant only one channel is connected to the common circuit. The principle of a tdm system is illustrated by Fig. 3.2 which shows the basic arrangement of a two-channel tdm system.

The two channels which are to share the common circuit are each connected to it via a *channel gate*. The channel gates are electronic switches which only permit the signal present on a channel to pass when opened by the application of a controlling pulse. Hence, if the controlling pulse is applied to gate 1 at time t_1 and not to gate 2, gate 1 will open for a time equal to the duration of the pulse but gate 2

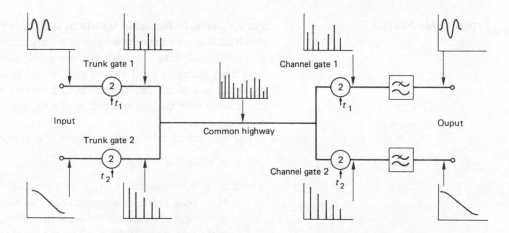

Fig. 3.2 A simple 2-channel tdm system. t_1 = a series of pulses occurring at fixed intervals. t_2 = a series of pulses occurring at the same periodicity as t_1 but commencing later by an amount equal to half the time interval (From B.T.E. Journal)

will remain closed. During this time, therefore, a pulse or sample of the amplitude of the signal waveform on channel 1 will be transmitted to line. At the end of the pulse, both gates are closed and no signal is transmitted to line. If now the controlling pulse is applied to gate 2 at a later time t_2, gate 2 will open and a sample of the signal waveform on channel 2 will be transmitted. Thus if the pulses applied to control the opening and shutting of gates 1 and 2 are repeated at regular intervals, a series of samples of the signal waveforms existing on the two channels will be transmitted.

At the receiving end of the system gates 1 and 2 are opened, by the application of control pulses, at those instants when the incoming waveform samples appropriate to their channel are being received. This requirement demands accurate *synchronization* between the controlling pulses applied to the gates 1, and also between the controlling pulses applied to the gates 2. If the time taken for signals to travel over the common circuit was zero then the system would require controlling pulses as shown, but since, in practice, the transmission time is not zero, the controlling pulses applied at the receiving end of the system must occur slightly later than the corresponding controlling pulses at the sending end. If the pulse synchronization is correct the waveform samples are directed to the correct channels at the receiving end. The received samples must then be converted back to the original waveform, i.e. demodulated. Provided the sampling rate, i.e. the number of controlling pulses per second, is at least equal to twice the highest frequency contained in the original waveform, correct demodulation can be achieved by merely passing the samples through a filter network that passes freely all frequencies lower than the sampling frequency.

Types of Modulation

For a signal to be frequency translated to another part of the frequency spectrum it is necessary for the signal to vary one of the characteristics of a sinusoidal wave — usually known as the *carrier wave* — whose frequency occupies the required part of the spectrum. The process by which one of the characteristics of a carrier wave is modified in accordance with the characteristics of a modulating signal is known as modulation.

The general expression for a sinusoidal carrier wave is

$$v = V \sin (\omega t + \theta) \tag{3.1}$$

where v = instantaneous voltage of the wave,
  V = peak value or amplitude of the wave,
  ω = angular velocity of the wave in radians/second; ω is related to the frequency of the wave by the expression $\omega = 2\pi f$ where f is the frequency in hertz,
  θ = the phase of the wave at the instant when $t = 0$.

For modulation is it necessary to cause one of the characteristics of the wave to be varied in accordance with the waveform of the modulating signal. Three possibilities exist:

(a) the amplitude V of the wave may be varied to give amplitude modulation,

(b) the frequency $\omega/2\pi$ may be modified to give frequency modulation,

(c) the phase θ may be the variable, giving phase modulation.

To illustrate, graphically, the difference between these three types of modulation consider the case in which the modulating signal is a positive-going pulse as in Fig. 3.3(a).

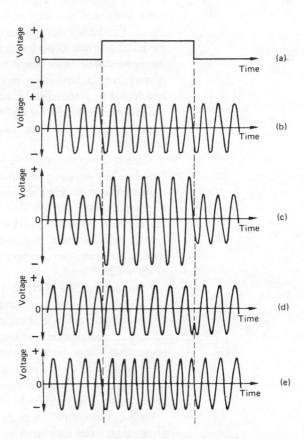

Fig. 3.3 Illustrating the difference between amplitude, frequency, and phase modulation
(a) Modulating signal
(b) Unmodulated carrier wave
(c) Amplitude-modulated wave
(d) Phase-modulated wave
(e) Frequency-modulated wave

In the case of the *amplitude-modulated* wave, the frequency of the carrier wave remains constant but its amplitude is suddenly increased when the leading edge of the modulating pulse occurs and then suddenly reduced to its original value when the pulse ends (Fig. 3.3(c)). The *phase-modulated* carrier wave has a constant amplitude and frequency, but experiences an abrupt change of phase at the instants corresponding to the beginning and end of the modulating signal pulse (Fig. 3.3(d)). Finally, the *frequency-modulated* carrier wave also has a constant amplitude, but the frequency of the wave is abruptly increased at the beginning of the modulating pulse and maintained at the new frequency until the pulse ends, when the frequency is suddenly decreased to its original value (Fig. 3.3(e)).

Amplitude modulation is employed for fdm telephony systems, radio broadcasting in the long, medium and short wavebands, the picture signal of 625 line television broadcasts and for various point-to-point and base-to-mobile radio-telephony systems.

Frequency modulation is used for data links, vhf radio broadcasting, the sound signal of 625 line television broadcasting, as well as for some base-to-mobile radio-telephony systems. Phase modulation is employed in its own right in data transmission links but otherwise it is mainly used as a stage in the production of a frequency-modulated signal. Frequency modulation may have an advantage over amplitude modulation in that its performance in the presence of interfering signals and noise can be much better, but it then suffers from the disadvantage of requiring a larger bandwidth. Because of the large bandwidth required for a frequency-modulated system the use of frequency modulation for broadcast and telephony circuits is restricted to very high frequencies where the bandwidth required can be more readily provided.

Systems using the principle of time-division multiplex are generally based on the sampling of the amplitude of the information signal at regular intervals, and the subsequent transmission of a pulse signal to represent each sample. Four main types of pulse modulation exist, namely pulse-amplitude modulation (pam), pulse-duration modulation (pdm), pulse-position modulation (ppm) and pulse-code modulation (pcm). The principle of each of the first three types of pulse modulation is shown in Fig. 3.4.

With pam, pulses of equal width and common spacing in time are employed and have their amplitudes varied in proportion to the instantaneous amplitude of the modulating signal. Pdm employs pulses of constant amplitude and common spacing between their leading edges, but whose width, or duration, is varied in accordance with the waveform of the modulating signal. Finally ppm employs constant amplitude and width pulses whose position (in time) is dependent upon the instantaneous amplitude of the modulating signal.

Each of these three methods of pulse modulation suffers from the disadvantage that distortion and noise is cumulative along the system — although the ppm system is much less affected than the other two in this way. The fourth type of pulse modulation, i.e. pulse code modulation (pcm), is a more sophisticated method which largely overcomes the effects of distortion and noise, but at the expense of added complexity and a wider bandwidth. In this system, pam signals are coded and then transmitted to line in digital form (p. 134).

Amplitude Modulation

If a carrier wave is amplitude modulated, the amplitude of the carrier is caused to vary in accordance with the instantaneous value of the

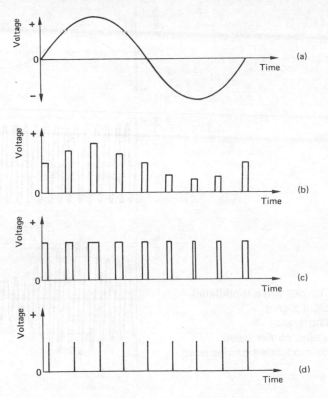

Fig. 3.4 Illustrating the difference between pam, pdm and ppm
(a) Modulating signal
(b) Pulse-amplitude modulation
(c) Pulse-duration modulation
(d) Pulse-position modulation

modulating signal. For example, consider the most simple case when the modulating signal is itself sinusoidal. The amplitude of the modulated wave must then vary in a sinusoidal manner as shown in Fig. 3.5. The difference in frequency between the carrier wave and the modulating signal will normally be much higher than is indicated by the diagram. In practice, the frequency difference would often be of the order of thousands of hertz.

The amplitude of the modulated carrier wave can be clearly seen to vary between a maximum value, which is greater than the amplitude of the unmodulated wave, and a minimum value, which is less than the amplitude of the unmodulated wave. The outline of the modulated carrier wave is known as the modulation *envelope*.

To convey intelligence, a signal containing at least two components at different frequencies is required and hence, in practice, a sinusoidal modulating signal is rarely used. However, whatever the waveform of the modulating signal the same principle still holds good; the envelope of the modulated wave must be the same as the waveform of the modulating signal. For example, Fig. 3.6 shows the envelopes of a carrier wave modulated by (*a*) a signal consisting of components at a fundamental frequency and its third harmonic, both components being in phase at time $t = 0$, (*b*) note C played on a cello organ pipe, and (*c*) a telegraphy signal produced by a teleprinter.

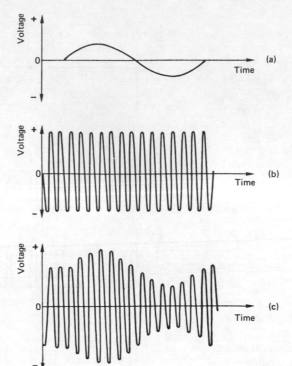

Fig. 3.5 A carrier wave modulated
by a sinusoidal signal
(a) Modulating signal
(b) Unmodulated carrier wave
(c) Amplitude-modulated carrier wave

The Frequencies in an Amplitude-modulated Wave

When a sinusoidal carrier wave is amplitude modulated, each frequency component in the modulating signal gives rise to two frequencies in the modulated wave, one below the carrier frequency and one above. When, for example, the modulating signal is a sinusoidal wave of frequency f_m the modulated carrier wave contains three frequencies:

    (*a*) the carrier frequency f_c;
    (*b*) the 'lower sidefrequency' $(f_c - f_m)$; and
    (*c*) the 'upper sidefrequency' $(f_c + f_m)$.

The two new frequencies are the upper and lower *sidefrequencies* $(f_c + f_m)$ and $(f_c - f_m)$ respectively, and these are equally spaced either side of the carrier frequency by an amount equal to the modulating signal frequency f_m. The frequency f_m of the modulating signal itself is *not* present.

    The bandwidth required to transmit an amplitude modulated carrier wave is equal to the difference between the highest frequency to be transmitted and the lowest. In the case of sinusoidal modulation and with $f_c > f_m$, the bandwidth required is given by

$$B = (f_c + f_m) - (f_c - f_m) = 2f_m$$

i.e. the required bandwidth is equal to twice the frequency of the modulating signal.

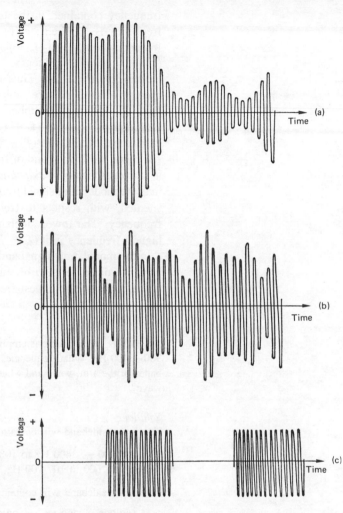

Fig. 3.6 Waveforms of a carrier wave modulated by
(a) a fundamental plus third harmonic
(b) a cello organ pipe playing C
(c) a telegraphy signal

EXAMPLE 3.1

A 100 kHz carrier is amplitude modulated by a sinusoidal tone of frequency 3000 Hz. Determine the frequencies contained in the amplitude-modulated wave and the bandwidth required for its transmission.

Solution
The frequencies contained in the modulated wave are

(1) The carrier frequency $f_c = 100\,000$ Hz   (*Ans.*)
(2) The lower sidefrequency $(f_c - f_m) = 100\,000 - 3000$
$\qquad\qquad\qquad\qquad\qquad\qquad = 97\,000$ Hz   (*Ans.*)
(3) The upper sidefrequency $(f_c + f_m) = 100\,000 + 3000$
$\qquad\qquad\qquad\qquad\qquad\qquad = 103\,000$ Hz   (*Ans.*)
$\qquad$ The bandwidth required $= 103\,000 - 97\,000 = 6000$ Hz   (*Ans.*)

If the modulating signal is non-sinusoidal it will contain components at a number of different frequencies; suppose that the highest frequency contained in the modulating signal is f_2 and the lowest frequency is f_1. Then the frequency f_2 will produce an upper sidefrequency component $(f_c + f_2)$ and a lower sidefrequency component $(f_c - f_2)$, while frequency f_1 will produce upper and lower sidefrequencies $(f_c \pm f_1)$. Thus the modulating signal will produce a number of lower sidefrequency components lying in the range $(f_c - f_2)$ to $(f_c - f_1)$ and a number of upper sidefrequency components in the range $(f_c + f_1)$ to $(f_c + f_2)$. The band of frequencies below the carrier frequency, i.e. $(f_c - f_2)$ to $(f_c - f_1)$, is known as the *lower sideband*, while the band of frequencies above the carrier frequency is known as the *upper sideband*. When the carrier frequency is higher than the modulating signal frequency the sidebands are symmetrically situated, with respect to frequency, on either side of the carrier frequency. The lower sideband is said to be *inverted* because the highest frequency in it $(f_c - f_1)$ corresponds to the lowest frequency f_1 in the modulating signal and vice versa. Similarly, the upper sideband is said to be *erect* because the lowest frequency in it $(f_c + f_1)$ corresponds to the lowest frequency f_1 in the modulating signal.

EXAMPLE 3.2

A 108 kHz carrier wave is amplitude modulated by a band of frequencies, 300–3400 Hz. What frequencies are contained in the upper and lower sidebands of the a.m. wave and what is the bandwidth required to transmit the wave?

Solution
The lower sideband will contain frequencies in the band

> 108 000 − 3400 Hz to 108 000 − 300 Hz,
> or 104 600 to 107 700 Hz   (*Ans.*)

The upper sideband will contain frequencies between

> 108 000 + 300 Hz to 108 000 + 3400 Hz,
> or 108 300 to 111 400 Hz   (*Ans.*)

The required bandwidth B is equal to the maximum frequency contained in the modulated wave minus the minimum frequency:

> $B = 111\ 400 - 104\ 600 = 6800$ Hz   (*Ans.*)

Note that the bandwidth is equal to twice the highest frequency contained in the modulating signal.

It is possible to confirm graphically the presence of a number of frequencies in an amplitude-modulated wave. Consider, for example, the case of a 10 000 Hz carrier wave and a 2000 Hz modulating signal. The upper sidefrequency will be 12 000 Hz and the lower sidefrequency will be 8000 Hz. The waveforms of the carrier, the upper sidefrequency and the lower sidefrequency components are shown in Figs. 3.7(a), (b) and (c) respectively. The carrier component can be seen to complete 10 cycles, the upper sidefrequency component 12

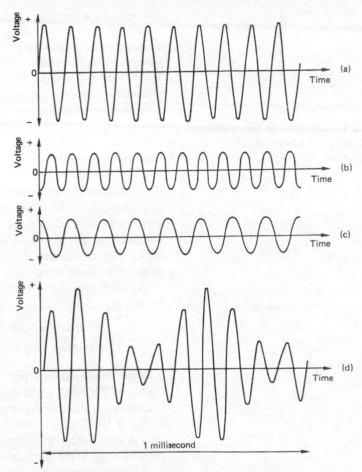

Fig. 3.7 Showing the formation of an amplitude-modulated wave by adding the components at the carrier and upper and lower sidefrequencies
(a) Carrier wave
(b) Upper sidefrequency
(c) Lower sidefrequency
(d) Amplitude-modulated carrier wave

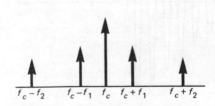

Fig. 3.8 Frequency spectrum of an amplitude-modulated wave

cycles and the lower sidefrequency component 8 cycles in the time of one millisecond. To obtain the waveform of the amplitude-modulated wave, which is composed of these three components, it is necessary to sum the instantaneous values of the three waves. It can be seen that the envelope of the modulated wave shows two complete variations in the time of one millisecond, i.e. it varies at the modulating frequency of 2000 Hz.

There are two ways in which the frequency spectrum occupied by a modulated carrier wave may be shown. Each component may be represented by an arrow drawn perpendicular to the frequency axis as shown in Fig. 3.8, where f_c, f_1 and f_2 have the same meanings as before. The lengths of the arrows are made proportional to the amplitude of the component they each represent. If many of the frequencies between f_1 and f_2 were present in the modulating signal, a large number of arrows would be required and the diagram would become impracticable. It is usual, particularly in connection with multi-channel carrier telephony systems, to represent the sidebands

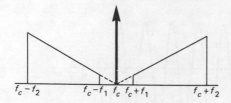

Fig. 3.9 A method of representing the sidebands of amplitude modulation

produced by a complex modulating signal by truncated triangles, in which the vertical ordinates are made proportional to the modulating frequency and no account is taken of amplitude — see Fig. 3.9. This method of representing sidebands gives an immediate indication of which sideband is erect and which is inverted. This is useful when considering systems employing more than one stage of modulation when the inverted sideband is not necessarily the lower sideband.

Modulation Depth

The envelope of an amplitude-modulated carrier wave varies in accordance with the waveform of the modulating signal and hence there must be a relationship between the maximum and minimum values of the modulated wave and the amplitude of the modulating signal. This relationship is expressed in terms of the modulation factor of the modulated wave.

The *modulation factor m* of an amplitude-modulated wave is defined by the expression

$$m = \frac{\text{maximum amplitude} - \text{minimum amplitude}}{\text{maximum amplitude} + \text{minimum amplitude}} \qquad (3.2)$$

When expressed as a percentage m is known as the modulation depth, or the depth of modulation or the percentage modulation.

Consider a sinusoidally modulated wave such as the wave shown in Fig. 3.10. Since the envelope of the modulated carrier wave must vary in accordance with the modulating signal, its maximum amplitude must be equal to the amplitude of the carrier wave plus the amplitude of the modulating signal, i.e. $(V_c + V_m)$. Similarly, the minimum amplitude of the modulated wave must be equal to $(V_c - V_m)$,

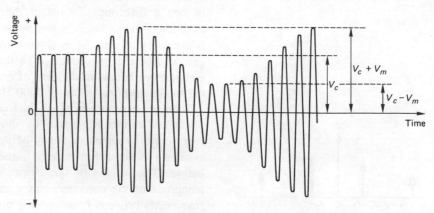

Fig. 3.10 Characteristics of an a.m. wave required to calculate modulation depth

where V_c is the amplitude of the carrier wave and V_m is the amplitude of the modulating signal.

For *sinusoidal modulation*, therefore, the modulation factor m becomes

$$m = \frac{(V_c + V_m) - (V_c - V_m)}{(V_c + V_m) + (V_c - V_m)} = \frac{V_m}{V_c} \qquad (3.3)$$

i.e. the modulation factor is equal to the ratio of the amplitude of the modulating signal to the amplitude of the carrier wave.

Also, for a sinusoidally modulated carrier wave, the amplitudes of the two sidefrequencies are the same and equal to $\frac{1}{2}m$ times the amplitude of the carrier wave.

EXAMPLE 3.3

Draw the waveform of an amplitude-modulated carrier wave that is sinusoidally modulated to a depth of 25%.

Solution

For a modulation depth of 25%, $m = 0.25$, and since the modulating signal is sinusoidal, $m = V_m/V_c$ or $V_m = 0.25 V_c$. Thus

the maximum amplitude of the modulated wave is $V_c + 0.25 V_c = 1.25 V_c$
and
the minimum amplitude is $V_c - 0.25 V_c = 0.75 V_c$

The required waveform is shown in Fig. 3.11.

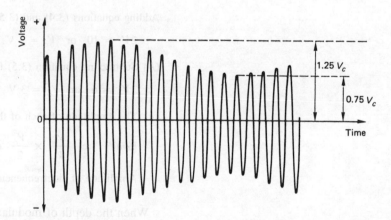

Fig. 3.11 Amplitude-modulated wave of modulation depth 25%

EXAMPLE 3.4

Determine the depth of modulation of the amplitude-modulated wave shown in Fig. 3.12.

Solution

The maximum voltage of the wave is 10 V and the minimum voltage is 5 V; hence the depth of modulation is

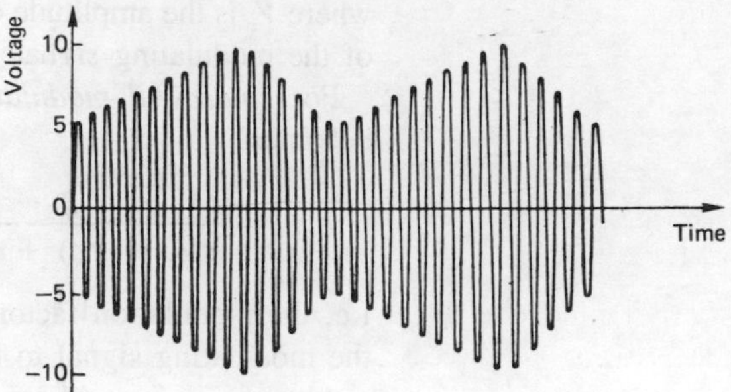

Fig. 3.12 Carrier wave amplitude modulated by a non-sinusoidal waveform

$$\frac{10 - 5}{10 + 5} \times 100\% = 33.3\% \quad (Ans.)$$

EXAMPLE 3.5

The envelope of a sinusoidally modulated carrier wave varies between a maximum value of 8 V and a minimum value of 2 V. Find (a) the amplitude of the carrier frequency component, (b) the amplitude of the modulating signal and (c) the amplitude of the two sidefrequencies.

Solution

(a) Maximum value $V_c + V_m = 8$ (3.4)

      Minimum value $V_c - V_m = 2$ (3.5)

Adding equations (3.4) and (3.5)

     $2V_c = 10$ or $V_c = 5$ V   (*Ans.*)

(b) Subtracting equation (3.5) from equation (3.4)

     $2V_m = 6$ or $V_m = 3$ V   (*Ans.*)

(c) The amplitude of each of the two sidefrequencies is equal to

$\frac{1}{2}mV_c$ i.e. $\dfrac{V_m}{V_c} \times \dfrac{V_c}{2}$ or $\frac{1}{2}V_m$

Amplitude of sidefrequencies = 3/2 or 1.5 V   (*Ans.*)

When the depth of modulation is 100% the modulation envelope varies between a maximum of $2V_c$ and a minimum of 0. If the depth of modulation is increased beyond this value the modulation envelope becomes distorted as shown in Fig. 3.13. The carrier is then said to be *over-modulated*.

Since the envelope of the modulated wave is no longer a replica of the modulating signal waveform, considerable distortion has taken place. This, of course, means that modulation depths greater than 100% are never employed in practice.

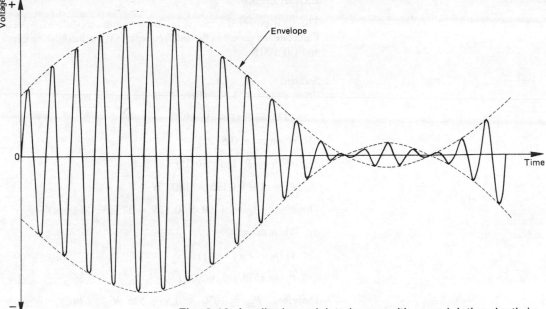

Fig. 3.13 Amplitude-modulated wave with a modulation depth in excess of 100%

Power of an Amplitude-modulated Wave

When an amplitude-modulated voltage wave is applied across a resistance R, each component frequency of the wave will dissipate power. The total power dissipated by the wave is the sum of the powers dissipated by the individual components. In the case of a sinusoidally modulated carrier wave, three components exist; namely, the carrier frequency, the upper sidefrequency, and lower sidefrequency.

The power P_c developed by each of the sidefrequency components is

$$P_c = \left(\frac{V_c}{\sqrt{2}}\right)^2 \cdot \frac{1}{R} = \frac{V_c^2}{2R} \text{ watts}$$

and the power developed by each of the sidefrequency components is

$$P_{LSF} = P_{USF} = \left(\frac{V_m}{2\sqrt{2}}\right)^2 \cdot \frac{1}{R} = \left(\frac{mV_c}{2\sqrt{2}}\right)^2 \cdot \frac{1}{R} = \frac{m^2 V_c^2}{8R} \text{ watts}$$

so that the total power P_t is

$$P_t = \frac{V_c^2}{2R} + \frac{m^2 V_c^2}{8R} + \frac{m^2 V_c^2}{8R}$$

or

$$P_t = \frac{V_c^2}{2R}\left(1 + \frac{m^2}{2}\right) \text{ watts} \tag{3.6}$$

EXAMPLE 3.6

The total power dissipated by an amplitude-modulated wave is 1575 W. Calculate the power in the sidefrequencies if the modulation depth is (i) 50% and (ii) 100%.

Solution
From equation (3.6), where P_c is the carrier power,

$$1575 = P_c(1 + \tfrac{1}{2}m^2)$$

(i) When $m = 0.5$

$$1575 = P_c(1 + \tfrac{1}{2}0.25)$$

$$P_c = 1575/1.125 = 1400 \text{ W}$$

Therefore $P_{SF} = 1400 \times 0.125 = 175 \text{ W}$ (*Ans.*)

(ii) When $m = 1$

$$1575 = P_c(1 + \tfrac{1}{2})$$

$$P_c = 1575/1.5 = 1050 \text{ W}$$

Therefore $P_{SF} = 1050 \times 0.5 = 525 \text{ W}$ (*Ans.*)

It is clear from this example that the power contained in the two sidefrequencies is only a small fraction of the total power, rising to a maximum when $m = 1$, that is only one-third of the total power. Since only the sidefrequencies carry information, amplitude modulation is not a very efficient system when considered on a power basis.

Single-sideband Operation

It is clear that an amplitude-modulated wave contains the intelligence represented by the modulating signal in both the upper sideband and the lower sideband. It is therefore unnecessary to transmit both sidebands. Furthermore, the carrier component is of constant amplitude and frequency and does not carry any information. It is possible (using a balanced modulator and a filter) to suppress both the carrier and one sideband in the transmitting equipment, and to transmit just the other sideband without any loss of information. This method of operation is known as *single-sideband suppressed-carrier* (ssb) working. The frequency spectrum graph of an ssb signal is shown in Fig. 3.14. Ssb operation of a communication system offers the following advantages over double-sideband (dsb) working.

(*a*) The bandwidth required for ssb transmission is only half as great as that required for dsb transmission carrying the same information. This allows more channels to be operated within the frequency spectrum provided by the transmission medium.

(*b*) The signal-to-noise ratio* at the receiving end of an ssb system is greater than that of a dsb system. The improvement is 9 DB for a depth of modulation of 100% and even more for modulation depths of less than 100%; some of this improvement is

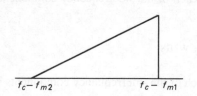

$f_c - f_{m2}$ $f_c - f_{m1}$

Fig. 3.14 Spectrum diagram of ssb signal

* Signal-to-noise ratio is discussed in Chapter 7.

the result of an increase in the ratio sideband power/total power of the transmitted output and the rest is because the necessary bandwidth is reduced by half (noise power is proportional to bandwidth).

(*c*) A dsb transmitter produces a power output (due to the transmitted carrier) at all times, whereas an ssb transmitter does not. A saving in the consumption of the d.c. power taken from the power supply is thus obtained, with an overall increase in transmitter efficiency.

(*d*) Radio waves are subject to a form of interference known as *selective fading*. When this is prevalent, considerable distortion of a dsb signal may occur, because the carrier component may fade below the level of the sidebands, so that the two sidebands beat together to produce a large number of unwanted frequencies. This cannot occur with an ssb system because the signal is demodulated against a locally generated carrier of constant amplitude.

(*e*) In multi-channel telephony line systems any non-linearity produces intermodulation products, many of which would lead to inter-channel crosstalk. Most non-linearity arises in the output stages of the line amplifiers since it is these that handle the signals of the greatest amplitude. Suppression of the carrier component reduces the signal levels to be handled by the amplifiers and this limits the effect of any non-linearity and hence also reduces crosstalk.

The disadvantage of ssb working is the need for more complex, and hence more costly, receiving equipment. The increased complexity is due to the need to reintroduce a carrier at the same frequency as the carrier originally suppressed at the transmitter. Any lack of synchronization between the suppressed and reintroduced carrier frequencies produces a shift in each component frequency of the demodulated signal. To preserve intelligibility the reinserted carrier must be within a few hertz of the original carrier frequency. The extra cost and complexity of operation are the reasons why ssb working is restricted to long-distance radio and line telephony systems and is not employed for domestic radio broadcasting.

Frequency Modulation

If a carrier wave is frequency-modulated, its frequency is made to vary in accordance with the instantaneous value of the modulating signal. The amount by which the carrier frequency deviates from its nominal value is proportional to the amplitude of the modulating signal, and the number of times per second the carrier deviates is equal to the modulating frequency. Figure 3.15(b) shows a frequency-modulated wave and Fig. 3.15(a) the corresponding modulating signal. Over the time interval 0 to t_1 the modulating signal voltage is zero and so the carrier is unmodulated. From t_1 to t_2 the modulating signal voltage is increasing in the positive direction and the carrier

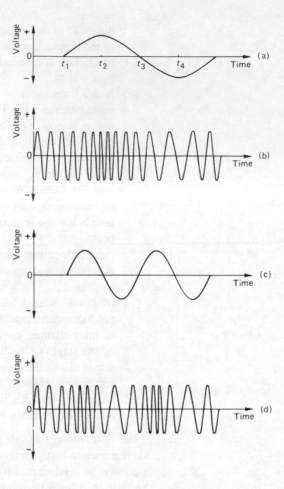

Fig. 3.15 Frequency-modulated waves

frequency increases to a maximum value at t_2. After time t_2 the modulating voltage falls towards zero, and the carrier frequency reduces in value until at time t_3 it has reached its unmodulated value. During the following negative half cycle of the modulating signal voltage, the carrier frequency is reduced below its unmodulated value. The carrier frequency has its minimum value when the modulating signal voltage reaches its peak negative value, i.e. at time t_4.

If the frequency of the modulating signal is increased (Fig. 3.15(c)), the number of times per second that the carrier frequency is varied about its mean, unmodulated value is increased in proportion. This is shown by Fig. 3.15(d). It should be noticed that the minimum and maximum values of the modulated carrier frequency are the same in both Figs 3.15(b) and (d); this is because the respective modulating signals have equal amplitudes.

Since the amplitude of a carrier wave is not changed when it is frequency modulated, there is no change in the power carried by the wave.

Frequency Deviation

The frequency deviation of a frequency-modulated wave is the amount by which the carrier frequency has been changed from its unmodulated value. Frequency deviation is proportional to the amplitude of the modulating signal voltage. There is no inherent maximum to the frequency deviation that can be achieved (unlike amplitude modulation where 100% modulation is the maximum amplitude deviation possible). Since the bandwidth occupied by a frequency-modulated wave is proportional to frequency deviation, a maximum value to the deviation permitted in a particular system must be arbitrarily chosen. The maximum value of frequency deviation that is allowed to occur in a particular system is known as the *rated system deviation*. Since frequency deviation is proportional to modulating signal amplitude, it follows that a maximum value, which depends on the sensitivity of the modulator, is also determined for the modulating voltage.

EXAMPLE 3.7

The rated system deviation in the BBC vhf sound broadcast system is 75 kHz. What will be the frequency deviation of the carrier if the modulating signal voltage is (i) 50% and (ii) 20% of the maximum value permitted?

Solution
 (i) Frequency deviation = 75 kHz × 0.5 = 37.5 kHz   (*Ans.*)
 (ii) Frequency deviation = 75 kHz × 0.2 = 15 kHz   (*Ans.*)

The term *frequency swing* is sometimes applied to a frequency-modulated wave. It refers to the maximum carrier frequency minus the minimum carrier frequency, i.e. the frequency swing is equal to twice the frequency deviation.

Modulation Index

When a carrier voltage is frequency modulated, its phase is also caused to vary. The peak phase deviation produced is equal to the ratio of the frequency deviation to the modulating frequency and is known as the modulation index of the modulated wave m_f. Thus

$$m_f = \frac{\text{frequency deviation}}{\text{modulating frequency}} \qquad (3.7)$$

EXAMPLE 3.8

A carrier is frequency modulated by a 10 kHz sinusoidal wave whose amplitude is one-half the maximum permitted value. If the rated system deviation is 50 kHz determine (i) the frequency deviation and (ii) the phase deviation, of the carrier produced.

Solution

(i) Frequency deviation = 50 kHz × 0.5 = 25 kHz   (*Ans.*)

(ii) $m_f = \dfrac{25 \text{ kHz}}{10 \text{ kHz}} = 2.5$ rad   (*Ans.*)

When both the frequency deviation and the modulating frequency are at their maximum permitted values, the modulation index is then known as the *deviation ratio D*, i.e.

$$D = \frac{\text{maximum frequency deviation}}{\text{maximum modulating frequency}} \tag{3.8}$$

The Frequencies in a Frequency-modulated Wave

When a carrier of frequency f_c is frequency modulated by a sinusoidal wave of frequency f_m, the resultant waveform contains components at a number of different frequencies. The modulated wave contains the following frequency components:

(*a*) the carrier frequency f_c
(*b*) first-order sidefrequencies $f_c \pm f_m$
(*c*) second-order sidefrequencies $f_c \pm 2f_m$
(*d*) third-order sidefrequencies $f_c + 3f_m$

and so on. The number of sidefrequencies present in a particular wave depends upon its modulation index, the larger the value of the modulation index the greater the number of sidefrequencies generated. The amplitudes of the various components, including the carrier itself, vary in a complicated manner as the modulation index is increased. Any component, again including the carrier, may have zero amplitude at a particular value of modulation index. The carrier is zero for $m_f = 2.405$, 5.32 and 8.654.

The bandwidth required to transmit a frequency-modulated wave is not, as might be expected, twice the frequency deviation of the carrier but, instead, is greater. The expression generally used to determine the bandwidth occupied by a frequency-modulated wave is given by equation (3.9), i.e.

$$\text{Bandwidth} = 2(f_d + f_m) \tag{3.9}$$

where f_d = frequency deviation of carrier, f_m = modulating frequency.

EXAMPLE 3.9

The BBC vhf frequency-modulated sound transmissions are operated with a rated system deviation of 75 kHz and a maximum modulating frequency of 15 kHz. Determine the required bandwidth. Compare this with the bandwidth required by an amplitude-modulation system that provides the same audio bandwidth.

Solution
From equation (3.9),

$$\text{Bandwidth} = 2(75 \times 10^3 + 15 \times 10^3) = 180 \text{ kHz} \qquad (Ans.)$$

If the same audio bandwidth were to be provided by an amplitude-modulated system, the necessary r.f. bandwidth would be $f_c \pm$ 15 kHz or 30 kHz. Clearly, in this case, frequency modulation is much more expensive in its use of the available frequency spectrum than is amplitude modulation. The signal-to-noise ratio at the output of a frequency-modulated system is proportional to the deviation ratio and, since D = maximum frequency deviation/maximum modulating frequency, also to the bandwidth occupied. This means that the signal-to-noise ratio can always be improved at the cost of an increased bandwidth; this is the reason why the BBC frequency-modulated sound broadcast stations are operated in the vhf band, since here the necessary wide bandwidth is available. With narrow-band frequency modulation (nbfm), the deviation ratio is small and the system offers little, if any, improvement in signal-to-noise ratio over the use of amplitude modulation. Nbfm is used for land mobile radio systems.

The Relative Merits of Frequency Modulation and DSB Amplitude Modulation

Frequency modulation offers the following advantages over the use of dsb amplitude modulation.

(*a*) The signal-to-noise ratio at the output of the f.m. receiver can be greater than that of a dsb amplitude-modulation receiver.
(*b*) The amplitude, and hence the power, of a frequency-modulated wave is constant, allowing more efficient transmitters to be built.
(*c*) An f.m. receiver possesses the ability to suppress the weaker of two signals simultaneously received at or near the same frequency. This effect is known as the *capture effect*.
(*d*) The dynamic range, i.e. the range of modulating signal amplitudes that can be transmitted, is much larger.

The disadvantage of frequency modulation is the (generally) wider bandwidth required.

Amplitude modulation is used for long, medium, and short wave-band broadcast transmissions, for the picture signal in television broadcasting, for long-distance radio-telephony, for vhf/uhf base-mobile systems, and for various ship and aeroplane radio services. Frequency modulation is used for vhf sound broadcasting, for the sound signal of uhf television broadcasts, for some base-mobile systems, for some ship/aero services, and for wideband radio-telephony systems.

Forms of both modulation methods are also used for the transmission of data signals and these will be considered in Chapter 8.

4 Carrier Frequencies and Bandwidths

A number of different methods of modulation are employed in telecommunication engineering to enable the best use to be made of the frequency spectrum available in a given transmission medium. For each channel in a communication system a *carrier frequency* must be chosen to position the channel in the required part of the frequency spectrum. To obtain the maximum utilization of the transmission medium and to minimize costs, the channel bandwidth must be as narrow as possible; it must, however, be wide enough to pass all the significant frequency components of the signal. The signals to be transmitted over the various types of communication system fall into one of five classes: (*a*) telephony, (*b*) telegraphy, (*c*) music, (*d*) television and (*e*) data transmission. This chapter will discuss the various factors that must be considered when determining the carrier frequencies and bandwidths for a particular system.

Line Communication Systems

Telephony

Telephone cables are capable of transmitting a band of frequencies well in excess of the normal speech frequency range and can therefore be used to carry single-sideband amplitude-modulated frequency-division multiplex telephony systems. A number of channels are made available by such a system, each channel being allocated a different carrier frequency. The range of carrier frequencies depends upon the number of channels provided by the system and the part of the frequency spectrum which the system is to occupy.

The number of channels that can be carried by a single cable pair is primarily determined by the attenuation (loss) of the cable at the highest frequency to be transmitted. The greater the cable attenuation the closer together must the line amplifiers be spaced in order to prevent the signal level falling below a predetermined value.

Most of the multi-channel fdm telephony systems in use in the United Kingdom consist of a suitable combination of CCITT*

*CCITT: International Consultative Committee for Telephony and Telegraphy.

12-channel carrier groups. By international agreement each channel in such a group must provide an audio bandwidth or *baseband* of 300–3400 Hz, which is sufficient to provide 'commercial-quality' speech. The CCITT 12-channel group will be considered in some detail in Chapter 11, but its basic principle is shown in Fig. 4.1. Each channel has a different carrier frequency that is amplitude modulated by the audio-frequency signal applied to the channel. The lower sidebands of each channel are selected by the channel filters and the composite signal is transmitted over the common line. At the receiving end, channel filters select the frequencies proper to each channel, and demodulators extract the original audio intelligence.

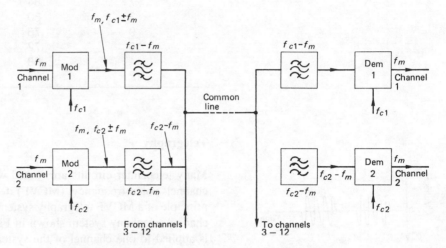

Fig. 4.1 Principle of a 12-channel telephony system

To keep inter-channel crosstalk to an acceptably low level, each filter is required to attenuate signals more than 600 Hz outside the 3.1 kHz channel bandwidth by at least 70 dB. To achieve this discrimination with a spacing of only 900 Hz between channels, *crystal filters* must be used. Crystal filters have very sharp attenuation/frequency characteristics, but are economic only at frequencies greater than about 60 kHz. For this reason the channel carrier frequencies start at 64 kHz for channel 12 and increase in 4 kHz steps to 108 kHz for channel 1. The lower sidebands of each channel are selected and so the bandwidth of the complete 12-channel system is 60.6–107.7 kHz (60.6 kHz = 64 − 3.4 kHz and 107.7 kHz = 108 − 0.3 kHz).

The channel carrier frequencies are specified by the CCITT and listed in Table 4.1 The table also gives details of the passband of each channel filter; it should be noted that the bandwidths correspond to an audio bandwidth of 300–3400 Hz.

The transmitted bandwidth is therefore 60.6–107.7 kHz, or approximately 60–108 kHz.

Table 4.1

Channel no.	Carrier frequency (kHz)	Channel filter passband (kHz)
1	108	104.6–107.7
2	104	100.6–103.7
3	100	96.6– 99.7
4	96	92.6– 95.7
5	92	88.6– 91.7
6	88	84.6– 87.7
7	84	80.6– 83.7
8	80	76.6– 79.7
9	76	72.6– 75.7
10	72	68.6– 71.7
11	68	64.6– 67.7
12	64	60.6– 63.7

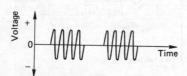

Fig. 4.2 Output waveform of a telegraph modulator

Telegraphy

Many teleprinter circuits are routed wholly or partly over a multi-channel voice-frequency (MCVF) telegraphy system. The basic principle of a MCVF telegraphy system is similar to that of the multi-channel telephony system shown in Fig. 4.1. The teleprinter signal is applied to one channel of the system and modulates the channel carrier frequency. The output from a channel send equipment consists of a series of pulses of carrier as shown in Fig. 4.2. Such a waveform is actually 100% double-sideband amplitude modulation.

The bandwidth required to transmit a teleprinter signal depends upon the characters sent, but the maximum bandwidth is demanded when alternate marks and spaces are transmitted, i.e. letters R and Y. The periodic time of one cycle of the waveforms for R and Y is 40 ms, and so the fundamental frequency of the waveform is 1000/40, or 25 Hz. A minimum channel bandwidth of 50 Hz is therefore required.

The choice of the carrier frequencies for the individual channels of a MCVF telegraphy system is determined primarily by the need to minimize inter-channel interference. The double-sideband outputs of the individual channels are combined and transmitted over the common path. Any non-linearity in the output/input characteristics of the line amplifiers, or other common equipment, will result in the presence of both *harmonic* and *intermodulation* products. For example, if f_1 and f_2 are two of the carrier frequencies, then the harmonic products are $2f_1$, $3f_1$, $4f_1$, etc., and $2f_2$, $3f_2$, $4f_2$, etc., and the intermodulation products are $f_1 \pm f_2$, $2f_1 \pm f_2$, $2f_2 \pm f_1$, etc. However, the products of greatest amplitude, and hence of the most importance, are $2f_1$, $2f_2$ and $f_1 \pm f_2$. To minimize inter-channel interference the channel carrier frequencies must be chosen to ensure that the important products do not fall within the passbands of the channel filters. This is achieved in practice by choosing odd harmonics of 60 Hz, starting with the seventh, as the channel carrier frequencies. Thus the channel carrier frequencies start at 420 Hz for channel 1 and increase in 120 Hz steps to 3180 Hz for channel 24.

Facsimile telegraphy is the transmission and reception of still pictures and diagrams.

Group 3 terminals currently dominate the fax market and these transmit the information at either 2400 b/s or 4800 b/s using FsK (Chapter 8). The standard frequencies are either 1300 ± 550 Hz or 1900 ± 550 Hz.

Radiocommunication Systems

General

The radiocommunication systems in use today may be divided into one of two main classes: firstly, *broadcasting* — both radio and television — and, secondly, *radio links* for providing point-to-point telephonic communication.

Table 4.2 gives the classification of the various frequency bands.

Table 4.2

Frequency band	Classification	Abbreviation
10–30 kHz	very low frequencies	vlf
30–300 kHz	low frequencies	lf
300–3000 kHz	medium frequencies	mf
3–30 MHz	high frequencies	hf
30–300 MHz	very high frequencies	vhf
300–3000 MHz	ultra high frequencies	uhf
3–30 GHz	super high frequencies	shf
30–300 GHz	extra high frequencies	ehf

Sound Broadcasting

A sound broadcasting system is one in which programmes of mixed entertainment, news and educational content are made available to a large number of people with radio receivers. At the lower frequencies a single transmitter can serve the whole country, but at higher frequencies the range of a particular station is much smaller and it becomes necessary to have a number of transmitting stations located in different parts of the country.

The BBC broadcast their Radio 1, 2, 3 and 4 programmes at a number of different frequencies in the mf and vhf bands and at one frequency in the lf band. In the mf band, amplitude-modulated dsb transmissions are used with carrier frequencies in the range 647–1546 kHz and having a bandwidth of approximately 9000 Hz. In Europe, by international agreement, carrier frequencies in the mf band are spaced at 9000 Hz intervals and, consequently, the highest audio frequency transmitted is in the region of 4500 Hz, i.e. a bandwidth of 9000 Hz.

The use of a 4.5 kHz audio bandwidth for mf broadcast transmissions means poor reproduction of music because many of the higher harmonics produced by musical instruments are suppressed. For reasonably high-quality reception of music, an audio bandwidth of at least 15 kHz is required. This occupies an r.f. bandwidth of 30 kHz, which cannot be accommodated in the congested mf band. High-quality broadcast transmissions are therefore provided in the vhf band using carrier frequencies in the range of 88.1–96.8 MHz. Vhf signals have a limited range of a hundred kilometres or so and so a number of stations are required within a fairly small area. The carrier frequencies allocated to the stations in a given area are about 200 kHz apart, to minimize inter-station interference. The wide carrier frequency spacing allows frequency modulation to be used; this gives a better signal/noise ratio than amplitude modulation but requires a much wider bandwidth (in the BBC vhf broadcast system an r.f. bandwidth of 180 kHz is necessary to provide the 15 kHz audio bandwidth).

Sound broadcast transmissions are also radiated in certain parts of the high-frequency band:

5.950–6.20 MHz, 7.1–7.3 MHz, 9.5–9.775 MHz,
11.7–11.975 MHz, 15.1–15.45 MHz, 17.7–17.9 MHz,
21.45–21.75 MHz and 25.6–26.1 MHz,

with a bandwidth of 10 kHz. Transmissions in these frequency bands are employed in Europe for international broadcast programmes as, for example, BBC programmes to Europe and to North Africa.

Television Broadcasting

Four television services are available to the British viewing public; two are provided by the BBC and a third by the various commercial television companies but transmitted by the Independent Broadcasting Authority (IBA). In addition, there is Channel 4. Colour and monochrome television signals are transmitted in the uhf band by both the BBC and the IBA. These programmes use vestigial sideband amplitude modulation for the picture signal, and frequency modulation for the sound signal. The vision carrier frequencies used are in the bands 471.25–575.25 and 615–847.25 MHz, providing a video bandwidth of 5.5 MHz. The sound carrier frequencies are spaced 6.0 MHz above the associated vision carrier and provide an audio bandwidth of 20 kHz.

Point-to-Point

British Telecom's (BT) hf radio links are used to route telephone calls to overseas destinations where line communication links via submarine cable or earth satellites are not available or have inadequate capacity. Different carrier frequencies are employed for each direction of transmission; and receiving stations are geographically situated well apart to minimize interference between signals passing in different directions. Radio links may be used to carry ordinary telephone conversations or to carry facsimile telegraphy or Press broadcasts. The majority of radio links have carrier frequencies in the hf band and generally use a form of ssb amplitude modulation that is known as *independent sideband*. Radio-telegraphy services, such as the Press Broadcast Service, which is provided for the use of news agencies, are also generally provided in the hf band.

A large number of frequency bands have been allocated to hf radio links; some of these, for example, are

3.5−3.9 MHz, 5.73−5.95 MHz, 9.04−9.5 MHz,
13.36−14 MHz, 21.75−21.87 MHz, 26.1−27.5 MHz.

The bandwidth per channel is 250 Hz−3 kHz.

Mobile Systems

Telephonic and telegraphic services to ships at sea are provided at a number of specified frequencies in the mf, hf, and vhf bands. Short-range telephony services, up to about 80 km in distance, are operated at frequencies in the vhf band, 156−163 MHz. Telephony services to ships at distances in excess of 80 km and less than about 1000 km are provided on specified carrier frequencies in the band 1.6−3.8 MHz. Ship-to-shore telegraphy services are operated, using the Morse code, in the band 405−525 kHz, except for some ships in northern seas which may use specified channels in the 1.6−3.8 MHz band. Longer distance services to ships at sea are operated at the following frequencies in the hf band; 4, 6, 8, 13, 17 and 22 MHz. 500 kHz (telegraphy) and 2182 kHz (telephony) are used as international distress and calling signals.

Many organizations, e.g. police, taxi-cab firms, and ambulances, use base-to-mobile systems to provide telephonic communication between a headquarters and several mobile units. Systems of this kind are operated in both the vhf band and the uhf band.

A considerable number of different frequency bands, of various widths, have been allocated to land, sea and air mobile services in the vhf and uhf bands. The complete list of frequencies is too long to include in this book and so Table 4.3 gives some examples.

A new mobile radio telephone service which provides telephonic communication between radio telephones in motor cars and the public

Table 4.3

Frequency band	Used by	Channel bandwidth (kHz)
71.5– 78.0	private land	12.5
80.0– 85.0	police and fire	12.5
85.0– 88.0	private land	12.5
97.0–102.0	police and fire	12.5
105.0–108.0	private land	12.5
108.0–136.0	aero	25
138.0–141.0	private land	12.5
146.0–148.0	police and fire	12.5
156.0–163.0	maritime	25
165.0–173.0	private land	12.5
425.0–449.5	private land	12.5
451.0–452.0	police and fire	25
453.0–462.5	private land	25
465.0–466.0	police and fire	25

switched telephone network now exists in the United Kingdom.

The frequency spectrum allocated to land mobile systems is subject to considerable demand and this new system of operation, known as *cellular radio*, has been introduced in the frequency band 860–960 MHz. Cellular radio provides very efficient usage of the allocated frequency spectrum since it permits frequent re-use of carrier frequencies.

Multi-channel Telephony/Television Links

In the uhf and shf bands, wideband radio systems, known as radio-relay systems, are an alternative to coaxial cable systems for the provision of multi-channel fdm systems between two points. In Great Britain microwave point-to-point links are provided in the following bands:

1.7–1.9 GHz (the 2 GHz spur band)
1.9–2.3 GHz (the 2 GHz main band)
3.7–4.2 GHz (the 4 GHz band)
5.85–6.425 GHz (the lower 6 GHz band)
6.425–7.11 GHz (the upper 6 GHz band)
10.7–11.7 GHz (the 11 GHz band).

The 2 GHz spur and main bands, the upper 6 GHz and the 11 GHz bands are split into a number of channels each of which is capable of handling 960 telephony channels or a television signal. The 2 GHz spur band is used to provide spur links within a city and for low-capacity links in remote areas. All other bands, including the 4 GHz and lower 6 GHz bands which are each divided into channels capable of accommodating 1800 channel telephony systems, are used for inter-city links, or links between a television studio and a television transmitter.

Many international telephony circuits are routed over multi-channel systems which are themselves routed over a satellite communication link. The internationally agreed frequency bands for satellite systems are

5.925−6.425 GHz for transmission from the earth to the satellite
3.7−4.2 GHz for transmission from the satellite to the earth.

Datacommunication

For reasons that will be discussed in Chapter 7 the d.c. data signal produced by a computer or by a peripheral data terminal cannot be directly transmitted over the telephone network. Instead, the data signal must be changed into a *voice frequency* (vf) signal using some form of *digital modulation*. If the signal is to be transmitted over the public switched telephone network the available bandwidth is only about 1400 Hz. At bit rates of up to 1200 bits/s a form of frequency modulation is employed that is commonly known as frequency shift modulation (or keying) (fsk). With fsk the data information is signalled by the transmission of either one of two different frequencies. These are 1300 Hz and either 1700 Hz or 2100 Hz. For higher bit rates more complex modulation methods are necessary but the signal can still be accommodated in the limited bandwidth offered by a telephone line.

The CCITT and the CCIR

To ensure compatibility between the telephone networks of different countries that play a part in a particular international telephone connection, it is necessary that carrier frequencies, bandwidths, noise levels and other parameters involved are standardized. The task of specifying the parameters of telecommunication systems destined for possible use in the international network has been given to the *International Telecommunication Union* (ITU). The ITU carries out its standardization work through two committees; the *International Consultative Committee for Radio* (CCIR) and the *Consultative Committee for Telephony and Telegraphy* (CCITT). The two committees meet at intervals to consider international telecommunication problems and policy, and include representatives from most, if not all, countries of the world. Sub-committees are set up to study particular telecommunication problems and to produce recommendations for the solution of those problems. The recommendations of the CCITT and CCIR are not mandatory but most equipments are manufactured to conform with those which are relevant. The application of CCITT/CCIR recommendations to equipment which will be used for purely national routes is a matter for the particular administration concerned, and for example, not all British Telecom's equipment conforms to recommended values. On the other hand, it is important that all equipment likely to be taken into use when an international connection is established should conform to all the relevant recommendations, otherwise many connections could have an inadequate transmission performance.

Among the CCITT/CCIR recommendations in current use are those covering the following topics.

(1) The carrier frequencies and the channel bandwidths to be used by the basic 12-channel telephony group.
(2) The way in which 12-channel groups can be combined together to form larger capacity coaxial systems.
(3) An international trunk switching plan.
(4) The frequency stabilities required for radio transmitters operating in different frequency bands.
(5) The allocation of frequencies to different services.

5 Basic Transmission Line Theory

Most telephone networks are divided into local lines, junctions and trunks. When analogue transmission techniques are employed, local lines are unamplified circuits that connect the individual telephone subscribers to their local telephone exchanges; junctions are two-wire circuits that may or may not be amplified and connect nearby telephone exchanges; and trunks are amplified four-wire circuits that interconnect more distant telephone exchanges.

Increasingly, digital transmission using *pulse code modulation* (pcm) is being employed to provide both junction and trunk circuits, and the intention of many telephone administrations is to convert their telephone networks into wholly digital systems as quickly as possible.

Most local lines and junctions employ pairs in audio-frequency telephone cables. Open-wire lines are frequently used for the final distribution of the local lines to the subscribers' premises and, rarely nowadays, for some junction circuits.

Most trunk circuits are routed wholly or partly over multi-channel fdm or tdm systems. These systems, in turn, are routed over either star-quad cables or coaxial telephony systems using either coaxial tubes or microwave radio-relay systems. Both two-wire lines and coaxial tubes are also used as feeders in radio stations to connect transmitters and/or receivers to aerials.

A transmission line consists of a pair of conductors separated from each other by a dielectric. Two main types of line exist: the two-wire, or twin, line, shown in Fig. 5.1(a) and the coaxial line shown at 5.1(b). The two-wire line may be either open-wire or a cable pair. The coaxial pair is nearly always operated with the outer conductor earthed since the outer then acts as an efficient screen at all operating frequencies, and is said to be *unbalanced*.

The conductors forming a pair have both *resistance* and *inductance* uniformly distributed along their length and uniformly distributed *capacitance* and *leakance* between them. These four quantities are known as the *primary coefficients* of a line.

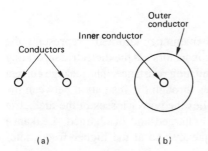

Fig. 5.1 (*a*) The two-wire pair (*b*) the coaxial pair

The Primary Coefficients of a Line

Resistance

The resistance R of a unit length of line, or *loop* resistance, is the sum of the resistances of the two conductors comprising a pair. The unit length of a line is the kilometre.

At zero frequency the resistance of a line is the d.c. resistance R_{dc} given by

$$R_{dc} = \frac{\rho_1}{a_1} + \frac{\rho_2}{a_2} \text{ ohms per loop kilometre} \tag{5.1}$$

where ρ_1 and r_2 are the resistivities of the two conductors, and a_1 and a_2 are their cross-sectional areas.

At a frequency of a few kilohertz or so, a phenomenon known as *skin effect* comes into play and causes current to flow only in a thin layer or 'skin' at the outer surface of the conductor. The higher the frequency the thinner this skin becomes and the smaller the cross-section of the conductor in which the current flows. Since the resistance is inversely proportional to the cross-sectional area of the 'effective' conductor, the *a.c. resistance* increases with increase in frequency. When skin effect is fully developed the a.c. resistance is proportional to the square root of the frequency, i.e.

$$R_{ac} = k_1\sqrt{f} \tag{5.2}$$

where k_1 is a constant.

Figure 5.2 shows how R_{ac} varies with frequency. Initially, little variation from the d.c. value is observed, but at higher frequencies the relationship given in equation (5.2) is true.

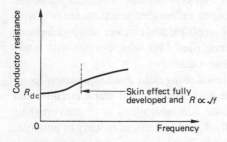

Fig. 5.2 Resistance/frequency characteristic of a transmission line

Inductance and Capacitance

The loop inductance L and the shunt capacitance C of a line, in henrys per loop kilometre and farads per kilometre respectively, are both more or less constant with change in frequency.

Leakance

The leakance G of a line in siemens per kilometre represents the leakage of current between the conductors via the dielectric separating them, and is the reciprocal of insulation resistance. The leakage current has two components: one passes through the insulation between the conductors, and the other supplies the power losses in the dielectric itself as the line capacitance is charged and discharged. Leakance increases with increase in frequency and at the higher frequencies it is directly proportional to frequency, i.e.

$$G = k_2 f \tag{5.3}$$

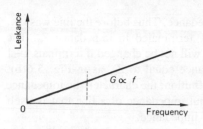

Fig. 5.3 Leakance/frequency characteristic of a transmission line

where k_2 is another constant. Figure 5.3 shows how the leakance of a pair of conductors varies with change in frequency.

A line can be represented by the network shown in Fig. 5.4. The line is considered to consist of a very large number of very short lengths, δl, of line connected in cascade. Each short section of line has a total shunt capacitance $C\delta l$ and shunt leakance $G\delta l$. The series inductance and resistance are really in both of the two conductors forming a pair but it is more convenient and customary to show them all to be in the upper conductor as in Fig. 5.4. Thus, the total series inductance and resistance per elemental length δl $L\delta l/2$ and $R\delta l/2$, respectively.

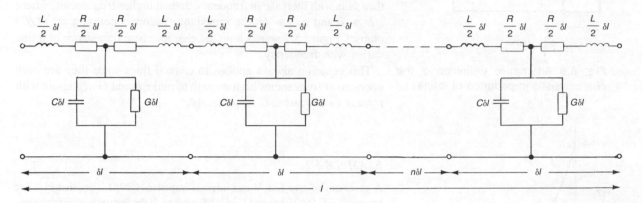

Fig. 5.4 Equivalent circuit of a line

The Secondary Coefficients of a Line

The secondary coefficients of a transmission line are its *characteristic impedance*, its *attenuation coefficient* and its *phase-change coefficient*.

Characteristic Impedance

The characteristic impedance Z_0 of a transmission line is the input impedance of an infinite length of that line. Figure 5.5 shows an infinite length of line; its input impedance is the ratio of the voltage V_s applied across the sending-end terminals to the current I_s flowing into the line, i.e.

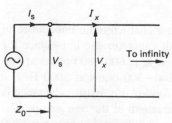

Fig. 5.5 Definition of the characteristic impedance of a line

$$Z_0 = \frac{V_s}{I_s} \text{ ohms} \tag{5.4}$$

Similarly, at any point x along the line, the ratio V_x/I_x is always equal to Z_0.

Suppose the line is now cut a finite distance from its sending-end terminals as shown in Fig. 5.6(a). The remainder of the line is still of infinite length and so the impedance measured at terminals 2−2

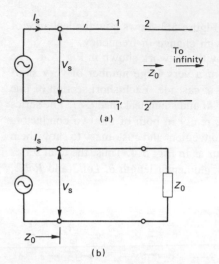

(a)

(b)

Z_0 →

Fig. 5.6 Alternative definition of the characteristic impedance of a line

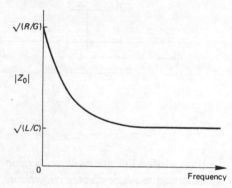

Fig. 5.7 Variation with frequency of the characteristic impedance of a line

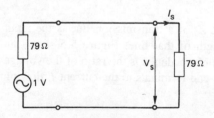

Fig. 5.8

*The full expression for the characteristic impedance of a line is

$$Z_0 = \sqrt{\frac{R + j\omega L}{G + j\omega C}} \text{ ohms}$$

is equal to the characteristic impedance. Thus before the line was cut, terminals 1−1 were effectively terminated in impedance Z_0. The conditions at the input terminals will not be changed if terminals 1−1 are closed in a physical impedance equal to Z_0, as in Fig. 5.6(b). This leads to a more practical definition: the characteristic impedance of a transmission line is the input impedance of a line that is itself terminated in the characteristic impedance.

A line that is terminated in its characteristic impedance is said to be *correctly terminated*.

The characteristic impedance of a line depends upon the values of the primary coefficients and also upon the frequency. At zero frequency the characteristic impedance is given by $\sqrt{(R/G)}$ ohms and then falls with increase in frequency until at higher frequencies, where $\omega L \gg R$ and $\omega C \gg G$, the impedance becomes constant at $\sqrt{(L/C)}$ ohms.* Figure 5.7 shows how the characteristic impedance of a line varies with frequency.

This equation always applies to coaxial lines since they are only operated at frequencies high enough to make R and G negligible with respect to ωL and ωC respectively.

EXAMPLE 5.1

A generator of e.m.f. 1 V and internal impedance 79 Ω is applied to a line having $L = 0.5$ mH/km and $C = 0.08$ μF/km. If the approximate expression, $Z_0 = \sqrt{(L/C)}$ ohms, for characteristic impedance may be assumed, calculate (*a*) the sending-end current, and (*b*) the sending-end voltage.

Solution

$$Z_0 = \sqrt{\frac{0.5 \times 10^{-3}}{0.08 \times 10^{-6}}} = 79 \text{ Ω}$$

Hence, referring to Fig. 5.8,

$$I_s = \frac{1}{79 + 79} = \frac{1}{158} \text{ A} \approx 6.33 \text{ mA} \quad (Ans. (a))$$

$$V_s = 79 I_s = \frac{79 \times 1}{158} = 0.5 \text{ V} \quad (Ans. (b))$$

Most practical coaxial cables have a characteristic impedance in the range 50−75 Ω. Typical values of characteristic impedance for unloaded two-wire audio-frequency cable is 600−800 ohms at low audio frequencies, falling to about 200−300 ohms at 3000 Hz.

When an audio-frequency cable is *loaded*, i.e. extra inductance is added at 1.828 km intervals along the length of the line (see p. 57), the value of the characteristic impedance is increased to about 1200 ohms.

Attentuation Coefficient

As a current or voltage is propagated along a line its amplitude is progressively reduced or *attenuated* because of losses in the line. These losses are of two types: first, conductor losses caused by I^2R power dissipation in the series resistance, and second, dielectric losses. If the current or voltage at the sending-end terminals of the line is I_s, or V_s, then the current, or voltage, at one kilometre distance along the line is $I_1 = I_s e^{-\alpha}$, or $V_1 = V_s e^{-\alpha}$, where e is the base of the natural logarithms (2.7183) and α is the *attenuation coefficient* of the line in nepers per kilometre. In the next kilometre distance the attenuation is the same, and thus the current I_2 at the end of this distance is

$$I_2 = I_1 e^{-\alpha} = I_s e^{-\alpha} e^{-\alpha} = I_s e^{-2\alpha}$$

If the line is l kilometres long, the received current and voltage are given, respectively, by

$$I_r = I_s e^{-\alpha l} \tag{5.5}$$

$$V_r = V_s e^{-\alpha l} \tag{5.6}$$

Thus both current and voltage waves decay exponentially as they are propagated along the line (Fig. 5.9).

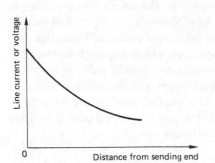

Fig. 5.9 Decay of current and voltage along a transmission line

$$I_x = I_s e^{-\alpha x} \quad \text{or} \quad V_x = V_s e^{-\alpha x}$$

Line current or voltage

0   Distance from sending end

EXAMPLE 5.2

A correctly terminated line has a characteristic impedance of 800 Ω, an attenuation coefficient of 0.3 Np and it is 4 km in length. Calculate the current in the load when the voltage applied across the sending terminals of the line is 2 V.

Solution
From equation (5.6)

$$V_r = 2e^{-0.3 \times 4} = 2e^{-1.2}$$
$$= 0.6024 \text{ V}$$
$$I_r = V_r/Z_0 = 0.6024/800 = 0.753 \text{ mA} \quad (Ans.)$$

The attenuation coefficient of a line may alternatively be quoted in decibels per kilometre and this is the usual practice in both the UK and USA.

EXAMPLE 5.3

Repeat Example 5.4 using decibels.

Solution
0.3 Np = 0.3 × 8.686 dB = 2.6058 dB and so the total line loss is

$$2.6058 \times 4 = 10.4232 \text{ dB}.$$

Therefore, $\quad 10.4232 = 20 \log_{10} 2/V_r$

or $\quad 10.4232/20 = 0.5212 = \log_{10} 2/V_r$

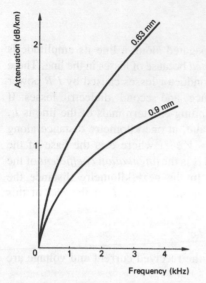

Fig. 5.10 Attenuation/frequency characteristics of audio-frequency star-quad cable

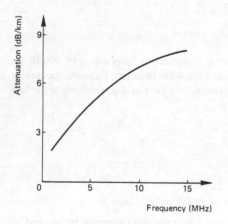

Fig. 5.11 Attenuation/frequency characteristics of a coaxial pair

Taking $a\log_{10}$ on both sides,

$$3.32 = 2/V_r \quad \text{or} \quad V_r = 2/3.32 = 0.6024 \text{ V}$$
$$I_r = 0.6024/800 = 0.753 \text{ mA} \quad (Ans.)$$

The attenuation of a cable is proportional to the length of the line. Thus if α is 3 dB/km at a particular frequency, the overall loss of a 2 km length of line is 6 dB, of a 4 km length of line is 12 dB, and so on.

Both the resistance and the conductance of a line increase with increase in frequency, and since these are the power-dissipating components the attenuation of the line increases with increase in frequency. Figure 5.10 shows how the attenuation coefficient of two types of a cable varies with frequency. The attenuation continues to increase at frequencies above those shown in the figure and may be as large as 25 dB/km at 1 MHz. Because of excessive losses like this, wideband telecommunication systems, which employ frequencies of some hundreds of kilohertz upwards, use *coaxial* cable as the transmission medium. The attenuation/frequency characteristic of a coaxial pair is shown in Fig. 5.11. The coaxial pair cannot be used at frequencies below about 60 kHz because the earthed outer conductor would then cease to act as an efficient screen.

EXAMPLE 5.4

A 6 km length of 0.63 mm star-quad cable is used as a data link. Determine the ratio of the attenuations experienced by the fundamental and the third and fifth harmonics of the data waveform when the bit rate is (*a*) 110 bits/s, (*b*) 1200 bits/s.

Solution
(*a*) When the bit rate is 110 bits/s the maximum fundamental frequency is 55 Hz, the third harmonic is 165 Hz, and the fifth harmonic is 265 Hz. From the curve given in Fig. 5.10 it can be seen that the values of the attenuation coefficient of the line at these three frequencies are small and difficult to determine, but clearly they are approximately equal.

(*b*) At 1200 bits/s the maximum fundamental frequency is 600 Hz, the third harmonic is 1800 Hz, and the fifth harmonic is 3000 Hz. From the curve,

$$\text{at} \quad 600 \text{ Hz} \quad \alpha \simeq 0.8 \text{ dB/km}$$
$$\text{at} \ 1800 \text{ Hz} \quad \alpha \simeq 1.4 \text{ dB/km}$$
$$\text{at} \ 3000 \text{ Hz} \quad \alpha \simeq 1.9 \text{ dB/km}$$

Thus the loss of a 6 km length of this cable is 4.8 dB at 600 Hz, 8.4 dB at 1800 Hz, and 11.4 dB at 3000 Hz.

4.8 dB is a voltage ratio of 1.74, 8.4 dB is a voltage ratio of 2.63, and 11.4 dB is a voltage ratio of 3.72. Therefore

$$\text{Ratio of attenuation} = 2.63/1.74 = 1.5 \quad \text{(third)}$$
$$\text{Ratio of attenuation} = 3.72/1.74 = 2.14 \quad \text{(fifth)}$$

When cables are connected in cascade their attenuation/frequency characteristics are *additive* as is shown by the following example.

EXAMPLE 5.5

The attenuation/frequency characteristics of four types of audio cable are given by Table 5.1.

Table 5.1

	Attenuation (dB/km)		
	800 Hz	1600 Hz	2000 Hz
Cable A	1.26	1.80	1.95
Cable B	1.13	1.61	1.74
Cable C	0.87	1.23	1.37
Cable D	0.59	0.81	0.89

Plot the overall attenuation/frequency characteristics of links consisting of the cascade connection of

(i) 2 km of cable A, 1 km of cable B, and 2 km of cable C,
(ii) 3 km of cable A, 2 km of cable C, 1.5 km of cable D, and 2.5 km of cable C.

Assume that there are no reflections at any of the cable junctions.

Solution
The required graphs are shown plotted in Fig. 5.12 and are obtained from the data given in Table 5.2.

Table 5.2

Link (i)	800 Hz	1600 Hz	2000 Hz
2 km of cable A	2.52	3.60	3.90
1 km of cable B	1.13	1.61	1.74
2 km of cable C	1.74	2.46	2.74
Total loss	5.39	7.67	8.38

Link (ii)			
3 km of cable A	3.78	5.40	5.85
2 km of cable C	1.74	2.46	2.74
1.5 km of cable D	0.89	1.22	1.34
2.5 km of cable C	2.18	3.08	3.43
Total loss	8.59	12.16	13.36

At frequencies where $\omega L \gg R$ and $\omega C \gg G$ the attenuation coefficient is given by

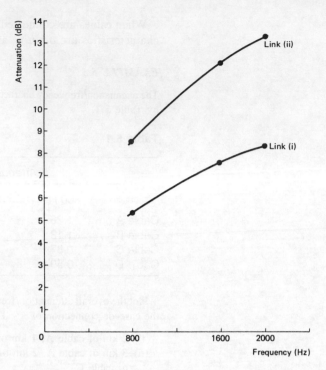

Fig. 5.12

$$\alpha = \frac{R}{2Z_0} + \frac{GZ_0}{2} \text{ nepers per kilometre} \qquad (5.7)$$

Usually, the second term is negligibly small and

$$\alpha = \frac{R}{2Z_0} \qquad (5.7(a))$$

EXAMPLE 5.6

A coaxial cable has a loss of 3.5 dB/km at 1 MHz. Calculate its loss at 4 MHz if (*a*) the dielectric loss is negligible, and (*b*) the dielectric loss is 10% of the total.

Solution

(*a*) Loss at 4 MHz = loss at 1 MHz $\times \sqrt{\dfrac{4 \times 10^6}{1 \times 10^6}}$

$$= 3.5 \times 2 = 7 \text{ dB/km} \qquad (Ans. \ (a))$$

(*b*) Dielectric loss at 1 MHz = 0.35 dB/km
Conductor loss at 1 MHz = 3.15 dB/km

$$\text{Loss at 4 MHz} = 3.15 \times \sqrt{\frac{4}{1}} + 0.35 \times \frac{4}{1}$$

$$= 6.3 + 1.4 = 7.7 \text{ dB/km} \qquad (Ans. \ (b))$$

Phase-change Coefficient

A current or voltage wave travels along a line with a finite velocity and so the current, or voltage, at the end of a kilometre length of line lags the current, or voltage, entering that length. The phase difference between the line currents, or voltages, at two points which are one kilometre apart is known as the *phase-change coefficient* β of the line. β is measured in radians per kilometre. In each kilometre distance of a line the same phase shift is introduced; consequently for a line l kilometre in length the received current will lag the sending-end current by βl radians.

EXAMPLE 5.7

A correctly terminated transmission line has $Z_0 = 500$ Ω, $\alpha = 1$ dB/km and $\beta = 30°$/km, and is 3 km long. A 500 Ω source, of e.m.f. 2 V, is applied to the sending-end terminals of the line. Calculate (a) the magnitude of the received current, and (b) its phase relative to the sending-end voltage.

Solution
Since the line is correctly terminated its input impedance is equal to its characteristic impedance of 500 Ω. Therefore

$$I_s = \frac{2}{500 + 500} = 2 \text{ mA}$$

Line loss $= 3 \times 1 = 3$ dB

The load and input impedances of the line are both 500 Ω and so use may be made of the expression

Attenuation in decibels $= 20 \log_{10}$ (current ratio)

Thus

$$3 = 20 \log_{10} \frac{2}{I_r}$$

$$\frac{2}{I_r} = \text{antilog}_{10} \frac{3}{20} = 0.1413 = \sqrt{2}$$

and

$$I_r = \frac{2}{\sqrt{2}} = \sqrt{2} \text{ mA} \quad (Ans. (a))$$

The phase shift introduced by the line is $3 \times 30° = 90°$; therefore

I_r lags V_s by $90°$ $\quad$ (*Ans.* (b))

Velocity of Propagation

Phase Velocity of Propagation

The phase velocity v_p of a line is the velocity with which a sinusoidal wave travels along that line. Any sinusoidal wave travels with a velocity of one wavelength per cycle. There are f cycles per second and so a wave travels with a velocity of λf metres per second, i.e.

$$v_p = \lambda f \text{ m/s} \tag{5.8}$$

where λ is the wavelength and f is the frequency of the sinusoidal wave.

In one wavelength a phase change of 2π radians occurs, and hence the phase change per metre is $2\pi/\lambda$ radians, and this is also equal to the phase-change coefficient. Thus

$$\beta = \frac{2\pi}{\lambda} \quad \text{or} \quad \lambda = \frac{2\pi}{\beta} \tag{5.9}$$

and

$$v_p = \frac{2\pi}{\beta} \times f = \frac{\omega}{\beta} \text{ m/s} \tag{5.10}$$

Group Velocity

Any repetitive, non-sinusoidal waveform contains components at a number of different frequencies, each of which will be propagated along a transmission line with a phase velocity given by Equation (5.10). For all these components to travel with the same velocity and arrive at the far end of the line at the same moment, it is necessary for the phase change coefficient β of the line to be a linear function of frequency, i.e. for ω/β to be a constant at all frequencies. It is only at radio frequencies that practical lines satisfy this requirement. At lower frequencies β varies with frequency in a non-linear manner. Figure 5.13 shows the β-frequency characteristic of a typical audio-frequency cable.

Suppose a signal consisting of a 1000 Hz fundamental plus 50% third harmonic is applied to a 10 km length of a pair in this cable. The fundamental frequency component will be propagated with a phase velocity

$$v_p = \frac{\omega}{\beta} = \frac{2\pi \times 1000}{0.075 \times 10^{-3}} \quad \text{or} \quad 83.78 \times 10^6 \text{ m/s}$$

and the component at the third harmonic will propagate with a phase velocity of

$$\frac{\omega}{\beta} = \frac{6\pi \times 1000}{0.155 \times 10^{-3}} \quad \text{or} \quad 121.61 \times 10^6 \text{ m/s}$$

This means that the harmonic component will arrive at the far end of the line t seconds before the fundamental arrives.

$$\text{Time } t = \frac{\text{Length of line}}{\text{Velocity difference}} = \frac{10^4}{(121.61 - 83.78) \times 10^6}$$

$$= 0.264 \text{ ms}$$

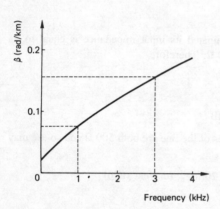

Fig. 5.13 Phase-change coefficient/frequency characteristic of a transmission line

This time is 0.264 times the periodic time of the fundamental frequency component and so the third harmonic component has a phase lead of 0.264 × 360° or approximately 95°. Figure 5.14(a) and (b) show, respectively, the resultant waveforms at the beginning and at the end of the line; it is evident that waveform distortion has taken place.

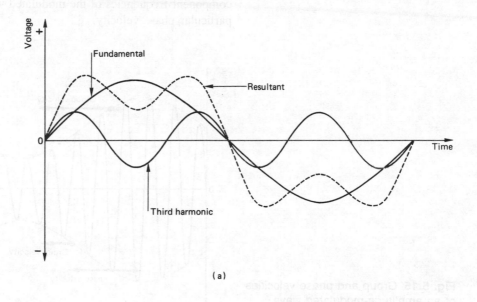

(a)

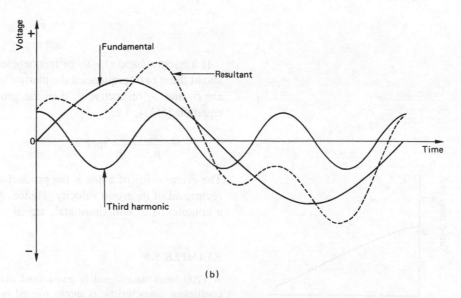

(b)

Fig. 5.14 Showing the envelope of a fundamental and its third harmonic (a) at the beginning and (b) at the end of a line in which the ratio ω/β is not constant

It is customary to consider the *group velocity* of a complex wave rather than the phase velocities of its individual frequency components. Group velocity is the velocity with which the *envelope* of the resultant waveform is propagated. Figure 5.15, for example, illustrates the meaning of the term group velocity when it is applied to the transmission of an amplitude-modulated wave over a line. The envelope travels at the group velocity, while the carrier, which is one of the component frequencies of the modulated wave, propagates with its particular phase velocity.

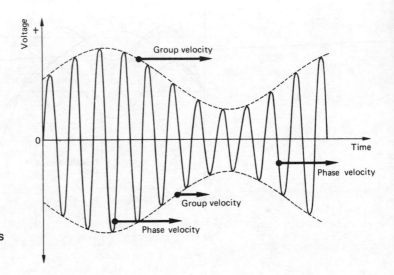

Fig. 5.15 Group and phase velocities of an amplitude-modulated wave

If a narrow band $\omega_2 - \omega_1$ of frequencies is transmitted over a line and at these two frequencies the phase change coefficients of the line are β_2 and β_1 respectively, then the group velocity V_g is given by equation (5.11), i.e.

$$V_g = \frac{\omega_2 - \omega_1}{\beta_2 - \beta_1} \text{ m/s} \tag{5.11}$$

The *group delay* of a line is the product of the length of the line and reciprocal of its group velocity. Hence, it is the propagation time of a complex, i.e. non-sinusoidal, signal.

EXAMPLE 5.8

A 1200 bits/s data signal is transmitted over a line whose phase-change coefficient characteristic is shown plotted in Fig. 5.13. The bandwidth of the transmitted signal is limited so that only frequencies up to the fifth harmonic are included. Determine (*a*) the group velocity of the signal, (*b*) the group delay if the line is 3 km in length.

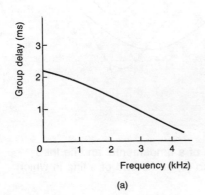

(a)

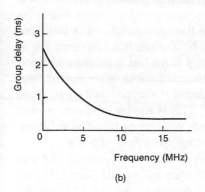

(b)

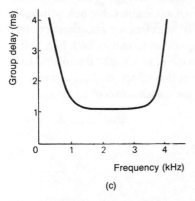

(c)

Fig. 5.16 Group-delay/frequency characteristics of (a) an audio-frequency cable, (b) a coaxial cable, and (c) a channel in a multi-channel telephony group

Loading of Cables

Solution

The maximum fundamental frequency of a 1200 bits/s data waveform is 600 Hz. The third and the fifth harmonics are 1800 and 3000 Hz. Thus the lowest and the highest frequencies to be transmitted are 600 Hz and 3000 Hz respectively. From the graph, the values of the phase-change coefficient at these two frequencies are 0.056×10^{-3} rad/km and 0.155×10^{-3} rad/km respectively.

$$V_g = \frac{2\pi(3000 - 600)}{0.155 - 0.056} = 152 \times 10^3 \text{ km/s}$$

The group delay is

$$\frac{1}{152 \times 10^3} \times 3 = 19.7 \ \mu s$$

The answer to this problem shows that, typically, the group delay per kilometre of an audio-frequency line is of the order of 7 μs per kilometre and this figure is negligibly small. Thus, it is only the high attenuation that stops long unloaded lines being used for digital data signals.

The group delay of a cable is proportional to the length of that cable. Thus if a cable has a group delay of 8 μs per kilometre at a particular frequency the group delay of a 10 km length will be 80 μs.

Three typical group delay/frequency characteristics are given in Fig. 5.16. Figure 5.16(a) shows clearly that the group delay on an audio-frequency cable decreases with increase in frequency; obviously it will also increase as the length of the line is increased. The group delay/frequency characteristic of a channel in a multi-channel telephony system increases at both low and high frequencies because of the presence of the channel filters. When the group delay of a circuit varies with frequency, the different components of a complex signal will not arrive at the end of the line at the same time and *group delay distortion* will be present. Because the group delay of a line is to some extent determined by its length, *relative* group delay is often quoted instead. The relative group delay is the difference between the propagation delays at a given frequency and those at a chosen reference frequency.

The attenuation of a cable is caused by its conductor I^2R losses and its dielectric V^2G losses. Conductor losses are the greater of the two, and indeed for coaxial cables (except the flexible type) the dielectric losses may be negligibly small. If the current flowing in a line for a given applied voltage could be reduced without a corresponding increase in the line resistance, a reduction in the attenuation could be achieved. At any point along a line the current is equal to the voltage at that point divided by the characteristic impedance of the line; this means that the line current could be reduced by increasing the characteristic impedance. The characteristic impedance of a line depends upon the values of each of the four primary coefficients and could be increased either by increasing the line inductance and/or the line resistance *or* by decreasing the line capacitance and/or leakance.

The minimum attenuation would be obtained if the condition $LG = RC$ could be achieved. In practice, it is the line inductance which is increased by the use of either *continuous loading* or *lumped loading*. With continuous loading each conductor is wrapped with a wire made from a high permeability material. This is an expensive process and it is therefore very rarely employed. With lumped loading inductors are inserted at regular intervals along the length of a line.

Three lump loadings are employed in the UK; these are 88 mH every 1.828 km, 120 mH every 1.828 km and 44 mH every 1.81 km, with 88 mH loading being the most common. Figure 5.17 shows the attenuation/frequency characteristic of a 0.9 mm cable pair with each value of loading inductance. It can be seen that the attenuation of the cable has been reduced at lower frequencies (compare with Fig. 5.10) but rises rapidly at some higher frequency. Clearly the loaded line acts like a low-pass filter. Because of this effect the loading must be removed if the line is to be used for the transmission of a pcm system.

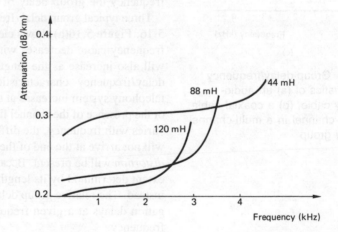

Fig. 5.17 Attenuation/frequency characteristics of loaded lines

The group delay/frequency characteristic of a loaded line depends upon the amount of loading and the length of the line. Figure 5.18(a) shows typical group delay/frequency characteristics of loaded lines of 16 km length and having loading inductances of 88 mH/1.828 km, 120 mH/1.828 km and 44 mH/1.81 km. Figure 5.18(b) shows group delay/frequency characteristics of different lengths of 88 mH/ 1.828 km loaded line. In these graphs delays of less than 5 μs have been ignored and taken as being equal to zero.

The standard loading used in the United Kingdom telephone network is 88 mH/1.828 km, and most junctions, except those converted for

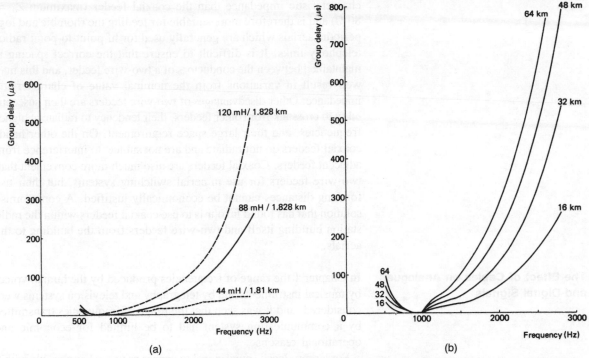

Fig. 5.18 (a) Group delay/frequency characteristics of loaded lines in 16 km lengths with different loadings
(b) Group delay/frequency characteristics of loaded lines in different lengths of 88 mH/1.828 km loading

pulse code modulation operation, are so loaded. The group delay of a loaded line rises rapidly at frequencies greater than about 1500 Hz and also at frequencies below about 300 Hz. The considerable variation of group delay with frequency means that a data waveform can only be directly transmitted over a line when the bit rate is very low, say 150 bits/s, unless the line is *very* short.

Group delay distortion in a coaxial cable system is negligibly small.

Use of Transmission Lines as Radio Station Feeders

In radio stations, both for transmission and reception, feeders are necessary to connect the station aerials to the radio transmitters or receivers. At a transmitting station the main requirement of a feeder is to transmit large amounts of power with utmost economy; in a receiving station the primary function of a feeder is to convey the signals picked up by an aerial to the receiver with minimum degradation in signal-to-noise ratio.

Both open-wire and coaxial lines are used as feeders, each having its particular merits. Open-wire feeders are cheaper to provide and to operate than coaxial feeders. The open-wire feeder has a higher characteristic impedance than the coaxial feeder (maximum $Z_0 \simeq$ 80 Ω) and is therefore more suitable for feeding the rhombic and log-periodic aerials which are generally used for hf point-to-point radio-telephony links. It is difficult to ensure that the correct spacing is maintained between the conductors of a two-wire feeder, and this may well result in variations from the nominal value of characteristic impedance. Other disadvantages of two-wire feeders are their susceptibility to crosstalk from other feeders, their tendency to radiate at higher frequencies, and their large space requirement. On the other hand, coaxial feeders do not radiate and are not subject to interference from adjacent feeders. Coaxial feeders are also much more convenient than two-wire feeders for use in aerial switching systems, but their use for long distances cannot be economically justified. A compromise solution that has found favour is to use coaxial feeders within the radio station building itself and two-wire feeders from the building to the aerials.

The Effect of Cables on Analogue and Digital Signals

In Chapter 1 the range of frequencies produced by the human voice, by musical instruments and by telegraphy and television systems were considered, and it was seen that the range of frequencies transmitted by a communication system had to be limited for economic and operational reasons.

For a reproduced sound signal to appear as natural as possible within the necessary bandwidth restrictions, it is essential that the original amplitude relationships between the fundamental component and the various harmonics are retained. As a signal is propagated along a transmission line it will be attenuated, and this attenuation is greater at the higher frequencies than at the lower. The effect of line attenuation is, therefore, to reduce the amplitudes of the harmonics relative to the fundamental and make the received sound seem unnatural. The effect on short telephone lines is slight and can be tolerated but longer telephone circuits and all music circuits are normally *equalized* (Chapter 9) to overcome this problem.

With telegraphy, data and television transmissions the signal waveform must be retained and this means that the various component frequencies must keep both their amplitude and phase relationships relative to one another and to the fundamental component. Thus, both amplitude/frequency and group delay/frequency distortion must be small. The bandwidth occupied by a telegraphy signal is so small that distortion is negligible on short circuits while all longer circuits are routed over a multi-channel system. The bandwidth occupied by a television signal is several megahertz and consequently both amplitude and group delay/frequency distortion are of importance. Lines for the transmission of television signals are generally fitted with both attenuation and group delay equalizers.

Mismatched Transmission Lines

Very often a transmission line is operated with a terminating impedance that is not equal to the characteristic impedance of the line.

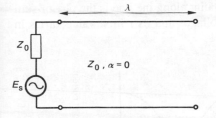

Fig. 5.19 Open-circuited loss-free line

When the impedance terminating a line is not equal to the characteristic impedance, the line is said to be *incorrectly terminated* or *mismatched*. Since the line is not matched, the load is not able to absorb all the incident power and so some of the power is *reflected* back towards the sending-end of the line.

Figure 5.19 shows a loss-free line whose output terminals are open circuited. The line has an electrical length of one wavelength and its input terminals are connected to a generator of e.m.f. E_s volts and impedance Z_0 ohms.

When the generator is first connected to the line, the input impedance of the line is equal to its characteristic impedance Z_0. An *incident* current of $E_s/2Z_0$ then flows into the line and an *incident* voltage of $E_s/2$ appears across the input terminals. These are, of course, the same values of sending-end current and voltage that flow into a correctly terminated line. The incident current and voltage waves propagate along the line, being phase-shifted as they travel. Since the electrical length of the line is one wavelength, the overall phase shift experienced is 360°

Since the output terminals of the line are open circuited, no current can flow between them. This means that *all* of the incident current must be *reflected* at the open circuit. The total current at the open circuit is the phasor sum of the incident and reflected currents, and since this must be zero the current must be reflected with 180° phase shift. The incident voltage is also totally reflected at the open circuit but with zero phase shift. The total voltage across the open-circuited terminals is twice the voltage that would exist if the line were correctly terminated. The reflected current and voltage waves propagate along the line towards its sending end, being phase shifted as they go. When the reflected waves reach the sending end, they are completely absorbed by the impedance of the matched source.

At any point along the line, the total current and voltage is the phasor sum of the incident and reflected currents and voltages. The way in which the r.m.s. line current varies with distance from the open circuit is shown by Fig. 5.20. The points at which maxima (*antinodes*) and minima (*nodes*) of current occur are always the same and do not vary with time. Because of this the waveform of Fig. 5.20 is said to be a *standing wave*.

At the open-circuited output terminals, the incident and reflected voltages are in phase and the total voltage is twice the incident voltage.

Fig. 5.20 The r.m.s. value of the total current along a loss-free open-circuited line

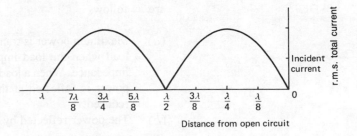

Distance from open circuit

Hence the total voltage at any point along the line is the phasor sum of the incident and reflected voltages and its r.m.s. value varies in the manner shown in Fig. 5.21.

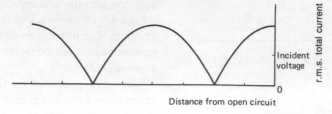

Fig. 5.21 The r.m.s. value of the total voltage along a loss-free open-circuited line

Distance from open circuit

Two things should be noted from Fig. 5.20 and Fig. 5.21. First, the *voltage standing-wave* pattern is displaced by $\lambda/4$ from the *current standing-wave* pattern, i.e. a current antinode occurs at the same point as a voltage node and vice versa. Second, the current and voltage values at the open circuit are repeated at $\lambda/2$ intervals along the length of the line; this remains true for any longer length of loss-free line.

When the output terminals of a loss-free line are short circuited, the conditions at the termination are reversed. There can be no voltage across the output terminals but the current flowing is twice the current that would flow in a matched load. This means that at the short circuit the incident current is totally reflected with zero phase shift and the incident voltage is totally reflected with 180° phase shift. Thus, Fig. 5.20 shows how the r.m.s. voltage on a short-circuited line varies with distance from the load, and Fig. 5.21 shows how the r.m.s. current varies. The standing waves of current and voltage on a $\lambda/4$ length of open-circuited line are easily obtained and are shown by Fig. 5.22.

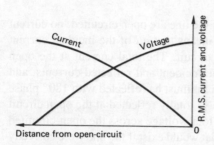

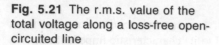

Fig. 5.22 Standing waves of current and voltage on a $\lambda/4$ open-circuited loss-free line

Standing Wave Ratio

An important parameter of any mismatched transmission line is its *voltage standing-wave ratio* or vswr. The vswr is the ratio of the maximum voltage on the line to the minimum voltage, i.e.

$$S = V_{\text{max}}/V_{\text{min}} \tag{5.12}$$

The presence of a standing wave on an aerial feeder is undesirable for several reasons and very often measures are taken to approach the matched condition and hence to minimize reflections. The reasons why standing waves on a feeder should be avoided if at all possible are as follows.

(*a*)   Maximum power is transferred from a transmission line to its load when the load impedance is equal to the characteristic impedance. When a load mismatch exists, some of the incident power is reflected at the load and the transfer efficiency is reduced.

(*b*)   The power reflected by a mismatched load will propagate, in the form of current and voltage waves, along the line towards its sending end. The waves will be attenuated as they travel and so the total line loss is increased.

(c) At a voltage maximum the line voltage may be anything up to twice as great as the incident voltage. For low-power feeders, such as those used in conjunction with radio receivers, the increased voltage will not matter. For a feeder connecting a high power radio transmitter to an aerial, however, the situation is quite different. Care must be taken to ensure that the maximum line voltage will not approach the breakdown voltage of the line's insulation. This means that for any given value of vswr there is a corresponding peak value for the incident voltage and hence for the maximum power that the feeder is able to transmit. A high vswr on a feeder can result in dangerously high voltages appearing at the voltage antinodes. Great care must then be taken by maintenance staff who are required to work on, or near to, the feeder system.

Measurement of Vswr

The vswr on a mismatched transmission line can be determined by measuring the maximum and minimum voltages that are present on the line. In practice, the measurement is generally carried out using an instrument known as a *standing-wave indicator*. Measurement of vswr not only shows up the presence of reflections on a line but it also offers a most convenient method of determining the nature of the load impedance.

The measurement procedure is as follows. The vswr is measured, using a detector, a galvonometer, or a vswr meter, and the distance in wavelengths from the load to the voltage minimum nearest to it is determined. The values obtained allow the magnitude and angle of the load impedance to be determined.

Methods of Matching

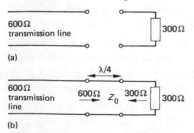

(a)

(b)

Fig. 5.23 (a) 600 Ω line and mismatched 300 Ω load, (b) use of a λ/4 section of low-loss line to obtain a matched system

A quarter-wavelength (λ/4) length of low-loss line possesses an important impedance-transforming property which is used for a wide variety of purposes at the higher frequencies.

A load impedance Z_L can be transformed into any desired value of input impedance Z_{in} by the suitable choice of the characteristic impedance Z_0 of a λ/4 length of low-loss line. The required value of the characteristic impedance is given by

$$Z_0 = \sqrt{(Z_L Z_{in})} \tag{5.13}$$

One common application of the *quarter-wave* (λ/4) *transformer* is the matching of a transmission line to a load impedance which is not equal to Z_0. Figure 5.23 shows a 600 Ω transmission line which is to be connected to a 300 Ω load impedance. If the line is directly

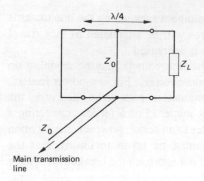

Fig. 5.24 Use of a λ/4 matching stub

connected to the load, reflections will occur and a voltage standing-wave pattern will appear on the line. To avoid this happening, the load impedance must be transformed into 600 Ω so that a matched system is obtained. A λ/4 matching section should be connected between the end of the 600 Ω line and the 300 Ω load as shown by Fig. 5.23. For the input impedance of the λ/4 section to be equal to 600 Ω, its characteristic impedance must be

$$Z_0 = \sqrt{(600 \times 300)} = 424.3 \ \Omega$$

Another method of using a λ/4 length of low-loss line as a matching device is shown by Fig. 5.24. The λ/4 section has one pair of terminals short-circuited and its other pair connected across the load impedance. The impedance of the λ/4 line will vary along its length from Z_L to zero and at some point it will be equal to the characteristic impedance of the main line. If the main line is connected to the λ/4 section at this point, it will be effectively correctly terminated and there will be no reflections.

Radiation from an Aerial

Whenever a current flows in a conductor, the conductor is surrounded by a magnetic field, the direction of which is determined by the direction of current flow. If the current changes, the magnetic field will change also. Now, a varying magnetic field *always* produces an electric field that exists *only* while the magnetic field continues to change. When the magnetic field is constant the electric field disappears. The direction of the electric field depends on whether the magnetic field is growing or collapsing and can be determined by the application of Lenz's law. Similarly, a changing electric field *always* produces a magnetic field; this means that a conductor carrying an alternating current is surrounded by continually changing magnetic and electric fields that are completely dependent on one another. Although a stationary electric field can exist without the presence of a magnetic field and vice versa, it is impossible for either field to exist separately when changing.

If a sinusoidal current is flowing in a conductor the electric and magnetic fields around the conductor will also attempt to vary sinusoidally. When the current reverses direction the magnetic field must first collapse into the conductor and *then* build up in the opposite direction. A finite time is required for a magnetic field and its associated electric field to collapse, however, and at frequencies above about 15 kHz not all the energy contained in the field has returned to the conductor before the current has started to increase in the opposite direction and create new electric and magnetic fields. The energy left outside the conductor cannot then return to it and instead, is propagated away from the conductor at the velocity of light (approximately 3×10^8 m/s), see Fig. 5.25. The amount of energy radiated from the conductor increases with increase in frequency, since more energy is then unable to return to the conductor.

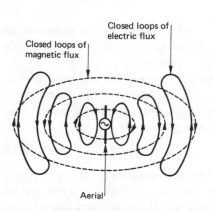

Fig. 5.25 Radiation from an aerial

The energy radiated from the conductor or aerial, known as the *radiated field*, is in the form of an *electromagnetic wave* in which there is a continual interchange of energy between the electric and magnetic fields. In an electromagnetic wave the electric and magnetic fields are at right angles to each other and they are mutually at right angles to the direction of propagation, as shown in Fig. 5.26 for a particular instant in time. The plane containing the electric field and the direction of propagation of the electromagnetic wave is known as the *plane of polarization* of the wave. For example, if the electric field is in the vertical plane, the magnetic field will be in the horizontal plane, and the wave is said to be vertically polarized. A vertically polarized wave will induce an e.m.f. in any *vertical* conductor that it passes, because its magnetic field will cut the conductor, but will have no effect on any horizontal conductor.

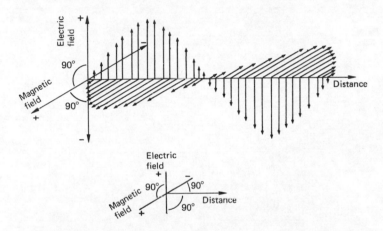

Fig. 5.26 The electromagnetic wave

In the immediate vicinity of an aerial the electric and magnetic fields are of greater magnitude and different relative phase than in the radiated field. This is because there is, in addition to the radiated field, an *induction field* near the aerial. The induction field represents energy that is not radiated away from the aerial, i.e. the energy that does succeed in returning to the conductor, and its magnitude diminishes inversely as the square of the distance from the aerial. The magnitude of the *radiated field* is proportional to the frequency of the wave and inversely proportional to the distance from the aerial. Near the aerial, the induction field is larger than the radiation field, but the radiation field is the larger at distances greater than $\lambda/2\pi$, where λ is the wavelength of the signal radiated from the aerial.

The amplitudes of the electric field E, and the magnetic field H, in an electromagnetic wave bear a constant relationship to each other. This relationship is known as the *impedance of free space* and is the ratio of the electric field strength to the magnetic field strength, i.e.

$$\text{Impedance of free space} = \frac{E \text{ (volts/metre)}}{H \text{ (ampere-turns/metre)}}$$

$$= 120\pi \text{ ohms} \tag{5.14}$$

$$= 377\ \Omega \tag{5.15}$$

It is customary to refer to the amplitude of a radio wave in terms of its electric field strength.

6 Noise

The output of a communication system, be it line or radio, will always contain some unwanted voltages or currents in addition to the desired signal. The unwanted output signal is known as *noise* and may have one or more of a number of different causes, each of which will be discussed in this chapter. For the signal received at the end of a system to be of use, the signal power must be greater than the noise power by an amount depending upon the nature of the signal. The ratio of the wanted signal power to the unwanted noise power is known as the *signal-to-noise ratio*. The signal level must never be allowed to fall below the value that gives the required minimum signal-to-noise ratio, because any gain introduced thereafter will increase the level of both noise and signal by the same amount and will not improve the signal-to-noise ratio.

Noise having a constant energy per unit bandwidth over a particular frequency band (uniform noise power density) is said to be *white noise*. Similarly, if the noise level decreases with increase in frequency at the rate of 3 dB/octave it is *pink*, while a noise power/frequency slope of 6 dB/octave gives *red* noise. (Note that the terms pink and red noise are not accepted by all workers in the field.) *Impulse noise* occurs in relatively high amplitude pulses that have a short duration compared with the time interval between the pulses.

The sources of noise are many; some noise is generated by various mechanisms within the equipment while further noise may by picked up by aerials and lines.

Sources of Noise

Thermal Agitation Noise

If the temperature of a conductor is increased from absolute zero (-273 °C), the atoms of the conductor begin to vibrate and some electrons are able to break away from their parent atoms and wander freely within the conductor. The amplitude of the atomic vibrations increases with increase in the temperature of the conductor and so the number of free electrons also increases with temperature. The free electrons wander in a random manner within the conductor, but at any particular instant more electrons are travelling in some directions than in others. The movement of an electron constitutes the flow of a minute current, and therefore the net current represented by the moving electrons fluctuates continuously in both magnitude and direction. Over a period that is long compared with the average time an electron travels in a particular direction, the total current is zero. The continuous flow of minute currents develops a random voltage across the conductor, and this unwanted voltage is known as *thermal agitation* noise or *resistance* noise.

The r.m.s. noise voltage produced by thermal agitation in a conductor is given by

$$V_n = \sqrt{(4kTBR)} \tag{6.1}$$

where k = Boltzmann's constant = 1.38×10^{-23} J/K

$\quad\quad\quad T$ = temperature of conductor in absolute degrees*

$\quad\quad\quad B$ = bandwidth (Hz) over which noise is measured, or of circuit at whose output the noise appears, whichever is the smaller. For most practical purposes, B can be taken as the 3 dB bandwidth of the circuit

$\quad\quad\quad R$ = resistance of circuit, ohms

Equation (6.1) may also be extended to find the noise voltage produced in an impedance; R is then the resistive component.

It is the bandwidth and not the frequency of operation that is important with regard to thermal agitation noise. Thus a wideband amplifier is noisier than a narrow-band amplifier whatever their operating frequencies may be.

EXAMPLE 6.1

Calculate the noise voltage produced in a 78 kΩ resistance in a 2 MHz bandwidth if the temperature is 20 °C.

Solution
From equation (6.1),

$$V_n = \sqrt{(4 \times 1.38 \times 10^{-23} \times 293 \times 2 \times 10^6 \times 78 \times 10^3)}$$
$$= 50.2 \ \mu V \quad (Ans.)$$

The thermal noise e.m.f. may be regarded as acting in series with the resistance R producing it. Maximum power transfer from a resistive source to a load occurs when the load resistance is equal to the resistance of the source. Consider a resistance R connected across another resistance of the same value that may be considered to be noiseless (Fig. 6.1). The noise power delivered to the load resistance is

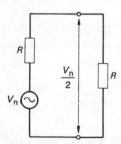

Fig. 6.1 Available noise power

$$P_a = \frac{(V_n/2)^2}{R} = \frac{4kTBR}{4R} = kTB \text{ watts} \tag{6.2}$$

Thus the maximum or *available noise power* that can be delivered by a resistance is independent of the value of that resistance but is proportional to *both* temperature *and* bandwidth. It is often convenient to note that, if the temperature is 290 K (17 °C), the available noise power is 4×10^{-15} W/MHz. Thermal agitation noise is *white*.

EXAMPLE 6.2

Calculate the available noise power from a resistance at a temperature of 17 °C over a 2 MHz bandwidth.

Solution
$$P_a = 2 \times 4 \times 10^{-5} = 8 \times 10^{-15} \text{ W} \qquad (Ans.)$$

The noise produced in most metallic resistors is purely the thermal agitation noise given by Equation (6.1). Most carbon resistors produce *current noise* in addition to this; current noise is caused by random variations in the contact resistance between carbon particles. The current noise e.m.f. increases with increase in both the current flowing in the resistor and in the resistance value, and is inversely proportional to frequency. At audio frequencies current noise may be larger than thermal agitation noise, but at radio frequencies thermal agitation noise predominates. The totak noise voltage V_{tn} produced in a carbon resistance is given by

$$V_{tn} = \sqrt{[(\text{thermal noise})^2 + (\text{current noise})^2]} \qquad (6.3)$$

Noise in Semiconductors

The output current of a transistor, fet or integrated circuit comprises a direct current determined by the d.c. operating conditions, plus a superimposed alternating current that is determined by the input signal current or voltage. Random fluctuations in these currents always exist, however, and may be considered to be the result of the superimposition of a noise current on the direct and signal currents.

Noise in Transistors

Thermal agitation noise. Thermal agitation noise is generated in the three regions of a bipolar transistor, but particularly in the base.

Shot noise. Shot noise in a transistor is caused by random fluctuations in the numbers of holes and electrons crossing each p-n junction. Since there are two p-n junctions in a transistor there are two sources of shot noise.

Partition noise. The input current to a transistor flows through the emitter to the base–emitter junction. After crossing the junction it divides between the collector and base terminals ($I_E = I_B + I_C$). This current division is also subject to random fluctuations and is thus another source of noise.

1/f Noise. Fluctuations in the conductivity of the semiconductor material produce a noise source which is inversely proportional to frequency. This noise, also known as current or excess noise, is usually negligible above about 10 kHz, and for some transistors above about 1 kHz.

Noise in FETs

Noise in a fet originates from three sources: shot noise generated by leakage currents in the gate-source p-n junction, thermal noise generated in the channel resistance, and 1/f noise caused by the random generation and recombination of charge carriers. The fet is inherently a lower noise device than is a transistor, although if a transistor is operated with a collector current of only a few microamperes its noise performance may be superior.

The reasons for the generally superior noise performance of a field-effect transistor are: (i) that its structure contains only one p-n junction as opposed to two in a bipolar transistor (this means that shot noise is less); and (ii) that the current flowing into the source can only flow out of the drain and this means that the fet is not subject to partition noise.

Noise in Line Systems

The noise output of a line communication system consists of noise generated in the transmission line itself and noise produced within the repeater stations along the route.

The noise arising in the transmission line is the sum of thermal agitation noise in the line resistance, noise due to faulty joints, interference picked up from nearby power lines or electric railways, and crosstalk from other pairs in the same cable. Thermal agitation noise has already been considered and noise caused by a faulty joint needs no discussion.

If a transmission line runs more or less parallel to a power line or an electric railway, it may have unwanted power-frequency voltages induced in it via inductive and/or capacitive couplings between the lines. Underground cables often have a metallic sheath, and this acts as a screen to reduce the magnitude of the unwanted voltages. Coaxial pairs are generally operated with their outer conductor earthed and are quite efficiently self-screened. This type of interference is minimized by keeping telecommunication cables spaced as far away from power lines as possible.

Crosstalk is a voltage appearing in one pair in a cable when a signal is applied to another pair. Any multi-pair cable will experience crosstalk between all its pairs to a greater or lesser extent. Crosstalk in a cable is caused by electrical couplings between the conductors; these couplings may be capacitive, magnetic or via insulation resistances.

The construction of a cable is designed to minimize crosstalk, and, when necessary, balancing the couplings between pairs at the end of each section of line can give a further reduction.

In a repeater station and a telephone exchange, noise voltages are introduced by thermal agitation in the equipment, faulty connections, inadequate smoothing of the power supplies, ic, transistor and fet noise, and crosstalk. Crosstalk occurs in the internal wiring because of capacitive and inductive couplings between wires. In the electronic equipment crosstalk is the result of electric couplings between inadequately screened components and couplings via the common power supplies. Crosstalk via the power supplies can occur because the current taken by each circuit flows in the common internal impedance of the power supply and develops a voltage across it. To minimize crosstalk in power supplies they are designed to have very low internal impedance.

A major cause of noise in fdm multi-channel systems is *intermodulation noise*. If a complex wave is applied to a device having a non-linear input/output characteristic, a number of new frequencies are produced and are present at the output. These new frequencies are equal to the sums and/or differences of the frequencies contained in the input signal. For example, if the input signal has components at frequencies f_1 and f_2 the output signal will contain components at

$$f_1 \pm f_2, \ 2f_1 \pm f_2, \ 2f_2 \pm f_1, \ 3f_1 \pm 2f_2, \text{ etc.}$$

in addition to the original frequencies f_1 and f_2. The number of extra frequencies thus produced can be very large; for example, if the input signal contains 100 different frequencies the number of *intermodulation products* runs into millions. In an fdm system (see p. 126) with all channels transmitting speech signals, the intermodulation products are beyond count and produce intermodulation noise at the output of the system. Intermodulation noise has components at all frequencies within the bandwidth of a channel and sounds very similar to thermal agitation noise. However, whereas thermal agitation noise is continuously produced, the magnitude of intermodulation noise depends upon the amplitude of the signals applied to the channels of the system.

Impulse noise, generated by a wide variety of sources, such as electric motors or switches may be picked up by wiring. Data circuits are more sensitive to noise and interference than speech circuits. For instance, short breaks in the transmission path of a millisecond or so would be unnoticed in a speech circuit, but could cause considerable error in a data circuit. Such breaks in the transmission path would probably be the result of poorly made joints in cable terminations and at distribution frames.

In addition to the various sources of noise previously mentioned, data circuits are also adversely affected by the following.

(*a*) Sudden changes in power level in fdm systems.
(*b*) Sudden changes in frequency in fdm systems.
(*c*) Impulse noise, such as dial impulses transmitted in one cable

pair appearing, as a result of crosstalk, in the data circuit. These are most troublesome if the bit duration is short in comparison with the time duration of the dial pulses. This source of noise will vary in intensity with the amount of telephone traffic and so it will vary with the time of day and the day of week.

(*d*) A major source of data noise in circuits routed via the public switched telephone network (pstn) is mechanical vibration of switches, particularly the wipers, in electro-mechanical telephone exchanges.

Because of (*d*), a data circuit routed over the pstn is noisier than one which uses a leased circuit.

Signal-to-Noise Ratio

The output of a communication system, whether line or radio, will always contain some unwanted voltages or currents in addition to the desired signal. The unwanted output signal is known as *noise* and may have one or more of a number of different causes. For the signal received at the end of a system to be of use, the signal power must be greater than the noise power by an amount depending upon the nature of the signal. The ratio of the wanted signal *power* to the unwanted noise *power* is known as the signal-to-noise ratio, i.e.

$$\text{Signal-to-noise ratio} = \frac{\text{Wanted signal power}}{\text{Unwanted noise power}} \qquad (6.4)$$

or

$$\text{Signal-to-noise ratio}$$
$$= 10 \log_{10} \left(\frac{\text{Wanted signal power}}{\text{Unwanted noise power}} \right) \text{dB} \qquad (6.5)$$

The signal-to-noise ratio required of a particular system depends upon the potential use of the signal and is generally determined by means of subjective tests. For example, a line music circuit may require a signal-to-noise ratio of 60 dB in order that the transmitted music may not be noticeably degraded, but a telephone circuit only requires about 35−40 dB. The required signal-to-noise ratio determines the spacing of the line amplifiers in a line communication system and of the relay stations in a microwave radio-relay link, and is a factor in the minimum transmitter power necessary in a radio system. For economic reasons, therefore, it is necessary to maximize the signal-to-noise ratio of a system by reducing the magnitudes of any noise sources as far as possible.

EXAMPLE 6.3

The signal voltage at the output of an amplifier is 1.2 V and the noise voltage that is unavoidably also present is 12 mV. Calculate the signal-to-noise ratio at the output of the amplifier.

Solution

Signal-to-noise ratio is the ratio of the wanted signal *power* to the unwanted noise *power*. Therefore

$$\text{Signal-to-noise ratio} = \frac{(1.2)^2}{(12 \times 10^{-3})^2} = 10\ 000 \text{ or } 40 \text{ dB} \qquad (Ans.)$$

Noise may arise from a number of different sources, some natural and some man-made (motor car ignition systems and electric motors, for example). Most naturally occurring noise sources produce a noise power which is directly proportional to bandwidth. This means that the signal-to-noise ratio at the output of an amplifier, a radio receiver, or some other kind of electronic, line or radio equipment is inversely proportional to the bandwidth of that equipment.

EXAMPLE 6.4

The signal-to-noise ratio at the output of a radio-frequency amplifier is 1000. What would be the signal-to-noise ratio if the bandwidth of the amplifier were doubled?

Solution

If the bandwidth of the amplifier is doubled the noise power at the output terminals will also be doubled. The signal output power is unchanged and so the signal-to-noise ratio will be reduced by half to 500.

EXAMPLE 6.5

An amplifier has a gain of 30 dB and generates a noise power, referred to its input terminals, of 3 μW. If the signal applied to the amplifier input is -10 dBm* with a signal-to-noise ratio of 20 dB, calculate the signal-to-noise ratio at the output of the amplifier.

Solution

Output signal level $= -10 + 30 = +20$ dBm
Input noise level $P_N = -10 - 20 = -30$ dBm

Therefore

$$-30 = 10 \log_{10} \left(\frac{P_N}{1 \times 10^{-3}} \right)$$

$$3 = \log_{10} \left(\frac{1 \times 10^{-3}}{P_N} \right)$$

Taking antilogs, $1000 = \dfrac{1 \times 10^{-3}}{P_N}$, so that

$$P_N = 1 \ \mu\text{W}$$

The total amplifier noise, referred to the input, is 3 μW, and hence the total input noise power is 4 μW. In dBm,

$$x = 10 \log_{10} \left(\frac{4 \times 10^{-6}}{1 \times 10^{-3}} \right) = 10 \log_{10} 4 \times 10^{-3} = -24 \text{ dBm}$$

*dBm: decibels relative to 1 milliwatt.

The output noise power is thus $-24 + 30 = +6$ dBm and the output signal-to-noise ratio is

$$20 - 6 = 14 \text{ dB} \quad (Ans.)$$

The noise generated in the line resistance will degrade the signal-to-noise ratio by an amount equal to its attenuation. If, therefore, a signal having a signal-to-noise ratio of 80 dB is applied to a line having a loss of 20 dB, the signal-to-noise ratio at the output of the line will be reduced to 60 dB. This figure will *not* be improved by amplifying the signal at a repeater station since *both* signal *and* noise will be amplified to the same extent. Instead, because of the various sources of repeater station noise, the signal-to-noise ratio will be further degraded.

In *any* analogue transmission system the signal-to-noise ratio must always progressively worsen as the length of the system is increased. This is to be contrasted with a digital system such as pulse code modulation in which noise is *not* cumulative with distance.

7 Digital Signals and their Transmission over Lines

In the past telephone networks have been primarily designed and planned for the transmission of analogue speech signals. In essence, this appears to be the simplest and most straightforward method of working since the speech signal need only be band-limited and amplified at suitable intervals along the line. Short lines, such as local lines connecting the user to his local telephone exchange, or junctions interconnecting nearby telephone exchanges, will require no processing at all.

A number of advantages can be gained, however, if the speech signal is converted before transmission to line into a representative binary-coded digital signal. The process employed to achieve this is known as *pulse code modulation* (pcm) and is considered in Chapter 12. The digital signal transmitted to line will be of the form shown in Fig. 7.1.

Fig. 7.1

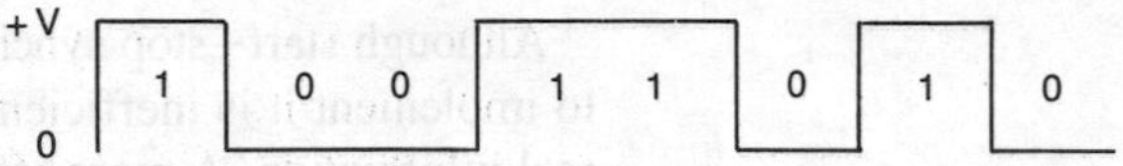

It can be seen to consist of a number of pulses of voltage which represent binary 1 and times when no pulse is sent to indicate binary 0. This is an example of a *single-current* or *unipolar* digital signal. Often, binary 0 is represented by a negative voltage pulse (see Fig. 7.2) to give a *double-current* or *bipolar* digital signal.

Fig. 7.2

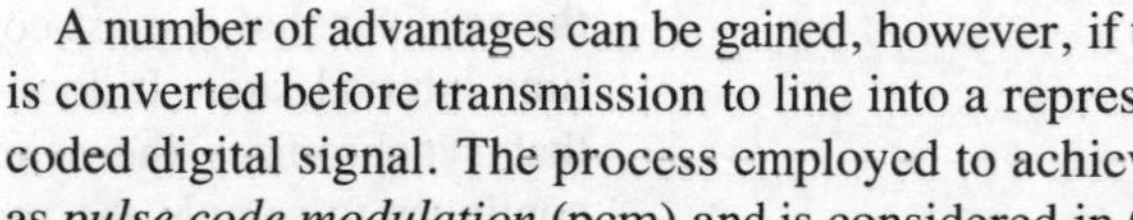

The d.c. data signals produced by a computer or a data terminal are of similar form and generally use International Alphabet 5 (ASCII) (see Appendix A). In this each character is represented by 7 bits each of which may be either a 1 or a 0. An example of ASCII is given in Fig. 7.3; it shows the characters B and R using *start—stop* synchronization. At the end of each character the signal always goes to the binary 1 state for a period of time at least equal to the time duration of a single bit. (Low-speed data signals may employ two stop bits.) The signal will remain in the stop state until the next character is to be transmitted; this is indicated by the state changing from binary 1 to binary 0 for a one-bit period before the bits representing the character begin.

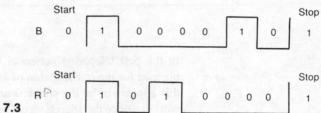

Fig. 7.3

The change in state from 1 to 0 at the start of each character specifies the start of a sampling period for the distant receiver; it indicates to the receiver that it must commence sampling the incoming bits one time interval later. It does this by starting the clock in the receiver that synchronizes the sampling instants. Thereafter, the receiver continues to sample the received bits during each of the following six time intervals (ideally in the middle of each). At the end of the last bit in a character the stop bit is received; this resets the receiver clock so that it is able to repeat the process when the next start bit arrives.

Although start—stop synchronization works and is relatively simple to implement it is inefficient since two in every nine bits convey no real information. A more efficient, although more complex, system is synchronous transmission of data. With this system start and stop bits are not employed. Instead, the receiver clock runs continuously and is kept in synchronism with the transmitter clock by synchronization pulses.

Digital signals cannot be transmitted over a telephone network, for other than very short distances, without some kind of processing. The resistance and capacitance of the line combine to distort the received waveform, and any line transformers, amplifiers and other equipment will remove the d.c. component (the average value). As a result, for other than very short distances, either *pulse regenerators* or *modems* must be employed.

The intention of most, if not all, telephone administrations is to convert their networks from analogue to wholly digital operation using pulse regenerators. The use of digital transmission instead of analogue transmission gives the following advantages.

(*a*) Better signal-to-noise ratio; this is because noise and interference are *not* cumulative with distance in a digital system but they are in an analogue system.

(*b*) Improved signalling.

(*c*) Multiplexing using tdm is simpler than multiplexing using fdm. The reasons for this are given in page 149.

(*d*) Digital switching is easier to implement.

(*e*) Different kinds of digital signal, e.g. data, telegraph, telephone speech and television, can be treated as identical signals during transmission and switching. This is the concept behind the new system of operation of telephone networks, known as Integrated Services Digital Network (isdn), that is just being introduced into networks throughout the world. Because of the economics involved the change from analogue to digital networks will take some time to implement.

The main disadvantage of digital operation is the wider bandwidths needed but the main reason for its employment being fairly recent is that the widespread introduction of many lsi integrated circuits has made complex circuit functions economic to carry out. An isdn is a single network that is able to transmit and switch a number of different communication services. An all-digital line network, based upon pcm telephony systems, is fully integrated with computer-controlled switching equipment in digital telephone exchanges. An isdn will be able to handle data, slow-scan television, teletext and telemetry signals as well as telephony speech signals with equal facility.

Single-current (Unipolar) and Double-current (Bipolar) Working

The digital signal that is transmitted to line can be either a *single-current* (unipolar) or a *double-current* (bipolar) signal. **Single-current operation** means that only one polarity of voltage is used, and data is signalled by the application of, and the removal of, the voltage applied to the sending end terminals of the line. When **double-current working** is used, two voltages are employed, one being positive with respect to earth and the other negative.

Both data and telegraph circuits can be operated in the *simplex*, the *half-duplex*, or the *full-duplex* modes. *Simplex* means that transmission of data is possible in only one direction over the link. A *half-duplex* circuit can be operated to transmit in either direction *but* only in one direction at a time. Lastly, a *full-duplex* circuit is capable of simultaneously transmitting data in both directions.

The principle of a *single-current simplex* circuit is illustrated by the basic teleprinter circuit shown by Fig. 7.4(a). When the keyboard of the printer is touched, the transmitter switch operates and connects

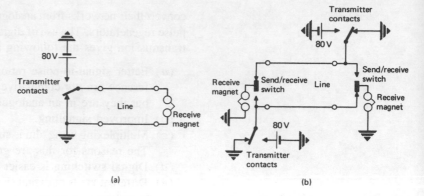

Fig. 7.4 (a) Single-current simplex circuit, (b) Single-current half-duplex circuit

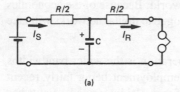

(a)

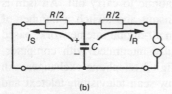

(b)

Fig. 7.5 Low-frequency equivalent circuit of a transmission line showing the currents flowing for single-current operation

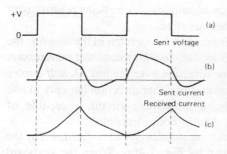

Fig. 7.6 Single-current waveforms

the +80 V battery to the line. Current then flows from the battery into the line and then through the receiving apparatus, before returning via the other conductor to the battery. The data is signalled by means of an interrupted d.c. voltage applied to the line. At zero and low frequencies the inductance and the leakance of the line both have negligible effect upon the signal and only the line's resistance and capacitance will affect the waveform of the received current.

Fig. 7.5 is an approximate representation of a low-frequency line which assumes that the total shunt capacitance is concentrated at the centre of the line. When the battery is connected to the sending end of the line, most of the current which flows into the line is used to charge the line capacitance. The current arriving at the far end of the line increases at a relatively slow rate and it is unable to reach its final *steady-state* value until the line capacitance has been fully charged. Once the line capacitance is fully charged, the sending-end current falls until it reaches the value needed to maintain the received current at its steady-state value. The time required for the received current to reach its steady-state value is proportional to the *total* capacitance C and resistance R of the line. Similarly, when the battery voltage is removed from the sending-end of the line and the terminal is earthed, the line capacitance will start to discharge at a rate determined by the time constant CRl^2, where l is the length of the line in kilometres. The discharge current is in the *opposite* direction to the original sending-end current, hence the sending-end current reverses its direction (see Fig. 7.5(b)). Some of the discharge current flows out of the receive end of the line in the *same* direction as the earlier current and so prolongs the receive current. Because of this the receive current does not fall as rapidly as the send current, neither does it reverse its direction. Figure 7.6 gives the waveforms of (a) the sent voltage, (b) the sent current, (c) the received current. The *risetime* of the output current is the time taken for the current to increase from 10% to 90% of its final value.

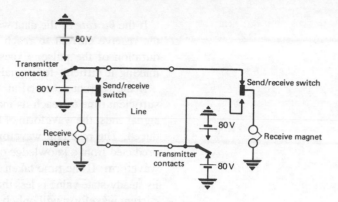

Fig. 7.7 Double-current half-duplex circuit

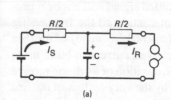

(a)

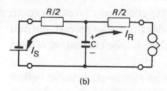

(b)

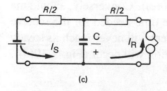

(c)

Fig. 7.8 Low-frequency equivalent circuit of a transmission line showing the currents flowing for double-current operation

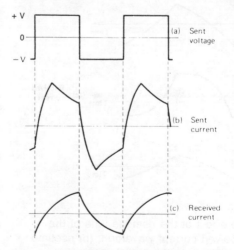

Fig. 7.9 Double-current waveforms

Figure 7.4(b) shows the arrangement of a *single-current half-duplex* telegraphy circuit; the operation is very similar to that just described except that transmission is possible in either direction.

Figure 7.7 shows the basic circuit of a *half-duplex double-current circuit*. The operation of the transmitter contacts reverses the polarity of the voltage supplied to the sending end of the line. Suppose a positive voltage, with respect to earth, is first applied to the line (Fig. 7.8(a)). When the line capacitance has been fully charged, the received current will reach its steady-state value after a time determined, as before, by the time constant CRl^2.

When the polarity of the sending-end voltage is first reversed (Fig. 7.8(b)), the line capacitance will start to discharge through both the transmitter and the receiver. The polarity of the reversed applied voltage is such that it acts in the same direction as the voltage developed across the line capacitance. As a result the capacitance of the line discharges much more rapidly than in the single-current case. This means that the receive current falls more rapidly. Once the line capacitance has been discharged, it is then charged in the opposite direction; most of the sending-end current is used to charge the line capacitance and thus the receive current increases, in the opposite direction to before, at a relatively slow rate, and it does not reach its steady-state value until the line capacitance has been fully charged (Fig. 7.8(c)).

Waveforms of current and voltage for a double-current circuit are given in Fig. 7.9. Clearly, the risetime and the falltime of the receive current waveform are smaller than in the single-current case. Further, the amplitude of the received current is greater.

Relative Merits of Single-current and Double-current Working

Single-current operation of a d.c. data circuit requires only one polarity voltage supply, whereas double-current operation requires the provision of both positive and negative power supplies.

If the *bit rate* of the data waveform is too high, the time taken for the receive current to reach its steady-state value may exceed the duration of the pulse. Excessive pulse distortion will then occur, making it difficult for signals to be reliably received.

If the current received at the end of a cable pair is not allowed sufficient time to reach its maximum value (steady-state) before the signal ends, the waveform of the transmitted signal will not be reproduced. The resultant waveform will not contain all the frequencies predicted from a knowledge of the fundamental frequency of the data waveform. If the time taken for the received current to build up to its steady-state value is less than the time duration of a bit, the receive current waveform will only be affected by the varying loss of the line at different frequencies. If the risetime of the receive current is greater than the bit length, distortion will occur. Some examples are given in Fig. 7.10. When the time taken for the received current to reach its steady-state value is equal to the bit duration, the receive current waveform is as shown by Fig. 7.10(b). Clearly, the current waveform is not rectangular.

If the receive current is able to reach its steady-state value before the trailing edge of the next voltage pulse occurs, its waveshape is approximately rectangular (see Fig. 7.10(c)). Conversely, should the pulse duration be much shorter than the time needed for the receive current to attain its steady state, the current will never reach its steady value and considerable waveform distortion will occur (Fig. 7.10(d)).

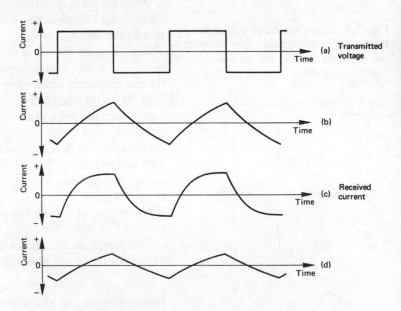

Fig. 7.10 (a) Showing the effect of the risetime time of the received current on the received current waveform; (b) risetime equal to bit duration; (c) risetime less than bit duration; (d) risetime greater than bit duration

Double-current operation of a data link increases the rate at which the received current rises towards its final value and so allows a higher bit rate to be employed. Double-current working also results in a larger-amplitude received current than does the single-current system for the same battery voltage. This gives more reliable operation.

A further disadvantage of the single-current method of operation is that any momentary break in the transmission path will not be detected as such but will be recorded as binary 0 and so produce an error in the received data.

The Arrival Curve

The *arrival curve* is a graph of the received current at the end of a line plotted against time for each positive or negative pulse taken separately, ignoring for simplicity any propagation delays (as was done in the previous figures). The arrival curve for a single pulse is shown by Fig. 7.11. The arrival curve can be used to construct the waveform of the received current for any input data waveform. This can be used to determine the time which is required for the received current to attain the value at which the receiver operates. The silent interval is equal to the time it takes the pulse to propagate over the line. It is the delay that occurs between a pulse being applied to one end of the line and the current starting to increase at the other end of the line.

The part of the arrival curve labelled as the transient is the time taken for the received current to reach its steady-state value. Each input voltage pulse is assumed to produce a current at the receiving end of the line corresponding to the arrival curve. The arrival curve for each input pulse is separately drawn and the waveform of the actual received current is then deduced by adding algebraically the individual arrival curves.

The sketches given in Fig. 7.12 apply this principle to, first, two pulses of differing time durations and, second, a square waveform, both signals being centred on 0 V.

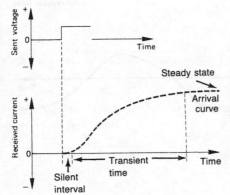

Fig. 7.11 Arrival curve for a single pulse

The Frequencies Contained in a Digital Waveform

The data fed into and out of a computer use ± 6 V pulses to represent the binary numbers 0 and 1, as shown by Fig. 7.13. Each signal element or *bit* has the same time duration in seconds as all the other bits and the number of bits transmitted per second is known as the *bit rate*. If, for example, the bit duration in Fig. 7.13 is 9.09 ms, the bit rate would be $1/9.09 \times 10^{-3}$ or 110 bits/s.

Figure 7.14 shows a sinusoidal voltage $v = V \sin \omega t$, of peak value V and frequency $\omega/2\pi$, and another, smaller, voltage at three times the frequency, i.e. $3\omega/2\pi$. The first voltage is the fundamental frequency and the other voltage is its third harmonic. At time $t = 0$, the two voltages are in phase with one another and the waveform produced by the summation of their instantaneous values is shown

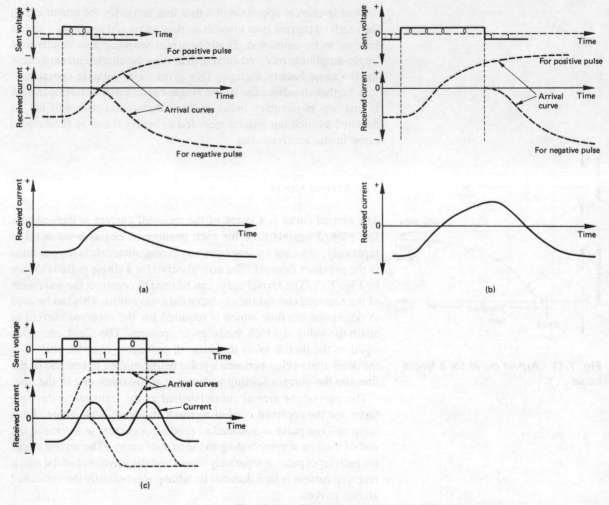

Fig. 7.12 Examples of the use of the arrival curve

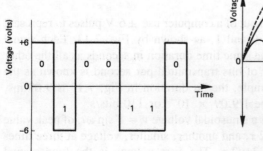

Fig. 7.13 Digital waveform

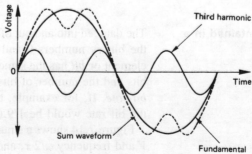

Fig. 7.14 Waveform produced by a fundamental and its third harmonic

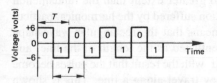

Fig. 7.15 Square digital waveform

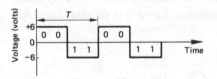

Fig. 7.16 A square digital waveform of lower bit rate

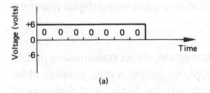

(a)

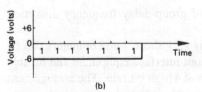

(b)

Fig. 7.17 Digital waveforms with zero fundamental frequency

as a broken line; clearly the resultant waveform is tending towards a rectangular shape.

If the fifth harmonic, also with zero phase angle at time $t = 0$, is added to the fundamental and the third harmonic, the resultant waveform is more nearly rectangular. Adding the seventh, ninth, etc. odd harmonics produces an even better approximation to the rectangular waveshape. If all the odd harmonics up to a very high order (theoretically infinity) are all included, a pulse train of square waveform is obtained (see Fig. 7.15).

The number of pulses occurring per second is known as the *pulse repetition frequency* (prf) and it is equal to the fundamental frequency contained in the pulse waveform. The *periodic time T* of the waveform is the reciprocal of the prf and is the time interval between the leading edges of consecutive pulses.

A digital waveform will only be square when it consists of alternate 0s and 1s as shown in Fig. 7.15. In the periodic time T, one 0 followed by one 1 occur, i.e. two bits. The fundamental frequency of this waveform is equal to $1/T$ and is one-half the number of bits per second. If the digital waveform consisted of alternate pairs of 0s and 1s as shown by Fig. 7.16, four bits occur in the periodic time T of the waveform and so the fundamental frequency of the waveform is now equal to one-quarter of the bit rate.

When the digital waveform is made up of a number of consecutive 0s or 1s (Fig. 7.17), the digital voltage is constant and so the frequency of the waveform is zero hertz.

The data transmitted over a link will include all sorts of combinations of 0s and 1s according to the information content, but the fundamental frequency produced will vary from a minimum of zero hertz to a maximum of one-half the bit rate. The more rapidly the digital waveform changes, or in other words the higher the bit rate, the higher will be the frequencies of its components. The minimum bandwidth that must be provided is equal to one-half the bit rate. If only this minimum bandwidth is provided, the digital waveform would lose its rectangular shape since none of its harmonics would be transmitted.

EXAMPLE 7.1

Determine the maximum fundamental frequency of a 147 bits/s data waveform. What other frequencies are present?

Solution

Maximum fundamental frequency = 147/2 = 73.5 Hz.

Other frequencies present are

(i) Third harmonic 220.5 Hz
(ii) Fifth harmonic 367.5 Hz, etc.

The Effect of Lines on Digital Signals

The attenuation of an audio-frequency cable increases with increase in frequency and so the various harmonics contained in a digital waveform will be attenuated to greater extent than the fundamental frequency. The greater attenuation suffered by the harmonics, particularly the higher-order ones, means that the rectangular waveshape will be lost. This effect is accentuated as the length of the line, and hence its attenuation is increased, with the result that the pulses become more and more rounded as they travel along a line. This is shown by Fig. 7.18. The higher the bit rate, the higher the fundamental frequency of the digital waveform and the shorter the length of line needed to reach the point at which satisfactory reception is impossible. The equipment at the receiving end of the line will sample each incoming bit at its midpoint and, provided it is able to reliably determine at any instant in time whether a 1 or a 0 is present, reception will be satisfactory.

Fig. 7.18 Effect of line attenuation on a transmitted digital waveform

Because of the effect of line attenuation, direct transmission of data waveforms over unamplified telephone cables is only possible at bit rates of up to 150 bits/s, and then only for fairly short distances of up to about 4 km.

At these bit rates, the effect of group-delay/frequency distortion is negligibly small.

If the distance between them is *very* short, two computers can be directly connected together without interface equipment and be able to communicate with one another at a high bit rate. The arrangement is shown by Fig. 7.19. Each computer is linked to its outside world by its input/output equipment; this in most cases consists of magnetic tape, floppy or hard discs, keyboard printers, and visual display units.

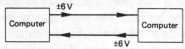

Fig. 7.19 Direct interconnection of two computers

The signals generated by digital equipment cannot be transmitted over a link set up via the analogue *public switched telephone network* (pstn) and/or over an amplified circuit because the d.c. component will be lost and the data waveform altered. (The d.c. component of a data waveform is the average value.) Further, if the bit rate is low the fundamental frequency and, perhaps, some of the lower-order harmonics may also not be transmitted. There are four reasons why the pstn cannot transmit the d.c. or very low frequency components of a data waveform.

(1) Transmission bridges in the telephone exchange switching equipment.
(2) Line matching transformers that are used in the audio-frequency

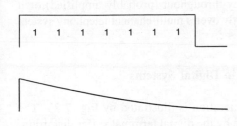

Fig. 7.20 Showing the effect of removing d.c. component

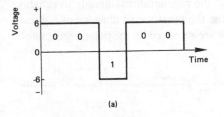

(a)

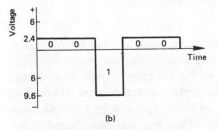

(b)

Fig. 7.21 Showing the effect of removing the d.c. component from a digital waveform: (a) with d.c. component, (b) without d.c. component

junction network to match different cables together and to match line amplifiers to cables.

(3) Line amplifiers incorporate capacitors and/or transformers as coupling components.

(4) Many links will be routed over one or more multi-channel telephony systems, or a radio link, and these neither provide a d.c. path nor pass frequencies below 300 Hz.

If the digital waveform consists of alternate 1s and 0s, its d.c. component will be zero (Fig. 7.15). At the other extreme when the digital waveform consists of a number of consecutive 1s, its d.c. component is −6 V; similarly, when consecutive 0s are sent, the d.c. value is +6 V. In either of these cases, the removal of the d.c. component would result in the effect shown in Fig. 7.20. Clearly, a loss of information is very likely.

Many other digital waveforms are also transmitted and, if the d.c. component is lost, the waveform will be altered.

Suppose, for example, the digital waveform is that shown by Fig. 7.21(a); the d.c. component, or average value, of this waveform is $3 \times \frac{6}{5}$ volts (since the 1 pulse cancels out one 0 pulse) or +3.6 V. If this component is removed from the waveform, the resulting waveform will vary from a positive voltage of $+6 - 3.6 = +2.4$ V to a negative voltage of $-6 - 3.6 = -9.6$ V, as shown in Fig. 7.21(b). The effects of line attenuation and noise will soon ensure that the receiving data terminal will be unable to reliably detect the binary 0 pulses.

There are two methods of overcoming these effects: either the digital signal can be converted to a voice-frequency signal, using digital modulation, in the normal commercial-quality speech bandwidth, or the pulses can be regenerated at regular intervals along the line.

The Use of Modems in Digital Systems

The modulation process is carried out by a piece of equipment known as a *modem* situated near the data source (Fig. 7.22). The digital signal is used to modulate a suitable carrier frequency and the modulated (voice-frequency) signal is transmitted over the circuit to the distant end. At the distant end another modem demodulates or detects the

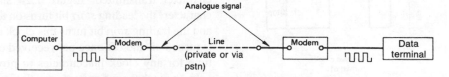

Fig. 7.22 Use of modems for data communication

incoming signal to recover the digital signal. The link may be a point-to-point circuit leased permanently from the telephone authority or it may be a dialled connection set up via the pstn. In either case, the link may be audio-frequency throughout (probably amplified) or it may be routed wholly or partly over a multi-channel telephony system or a radio link.

The Use of Regenerators in Digital Systems

The principle of pulse regeneration is illustrated by Fig. 7.23. The pulses transmitted to the line by the digital terminal suffer distortion, because of the combined effects of line attenuation and group-delay/frequency distortion and noise, but they are recreated as originally produced by each line pulse regenerator. The data waveform is reconstituted without error provided the pulse waveform has not been degraded to such an extent that the regenerator is unable to reliably determine at each instant in time the presence or absence of a pulse. Effectively, noise and distortion are removed by the pulse regenerator.

Fig. 7.23 Use of pulse regenerators

Synchronization

When two digital terminals are linked together and interchanging information, the terminal receiving information must be *synchronized*. *Synchronization* is essential so that the receiver will, at all times, sample each incoming bit at the correct instant in time. Otherwise, the possibility exists that one, or more, bits may be lost with catastrophic effects on the accuracy of the received data. Suppose, for example, that the three denary numbers 13, 17 and 15 are transmitted using the binary code. The transmitted data is then 011011000101111 (printed in the order of transmission, left-hand side first). If, because of a lack of synchronization, the initial bit is missed, the received data would become 11011000101111 or 27, 2 and either 31 or 30 (depending on whether the next bit to appear is a 1 or a 0).

There are two main methods of synchronization used in data networks known as *anisochronous* and *isochronous*. An anisochronous data system, used at low bit rates, inserts synchronizing bits into each character transmitted. Figure 7.24 shows the example of a 7-bit character; the leading *start* bit turns on the clock in the distant receiver and the trailing *stop* bit turns this clock off. Because the receive clock is turned on and off once per received character, there is insufficient time for any clock inaccuracies to produce a significant error.

In an isochronous system, the timing of both the receiver and the transmitter is controlled by a clock in *either* the transmitting data

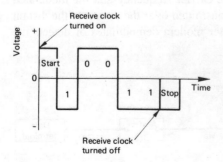

Fig. 7.24 Start—stop synchronization

terminal *or* its associated modem. The receiving modem and data terminal derive their timing information from the incoming data itself and/or synchronization bits inserted into the data stream. In one case, for example, the receiver compares the incoming bits with the receive clock, and if need be adjusts the clock to minimize any error. The synchronizing bits are far fewer than those required for an aniso-chronous system and so the system is more efficient in its transfer of actual data. Isochronous systems work at bit rates of 1200 bits/s and upwards.

Signalling Systems

Clearly, there is a need for signalling systems in a telephone network to make it possible for a customer, or an operator, to establish a connection and, later, to clear the connection.

The most common method of signalling in use is *loop-disconnect*, in which the action of a telephone dial, or equivalent key-pad, is to break, and make, the line loop to the local telephone exchange. The loop is broken a number of times that corresponds to the number dialled. Thus, if number 3 is dialled, three break and make pulses are generated. In the UK the dial break period is set at 67 ms and the make period at 33 ms; this means that 10 pulses (number 0) occupy one second. For signalling over distances up to about ten miles loop-disconnect or double-current d.c. systems can be employed. For longer lines it is often necessary to employ pulse regenerators to reshape the signalling pulses.

Longer distance circuits may employ amplifiers to overcome the losses of the line and these are unable to pass d.c. signals. The d.c. signals must then bypass each amplifier by making use of the phantom circuits. However, the distance over which it is possible to send signals is still limited. For all longer distance analogue circuits it is therefore necessary to use voice-frequency signals or *vf tones*. Vf signalling tones generally lie within the normal speech bandwidth. A number of different vf signalling systems have been employed and some UK examples are given below.

Single-frequency 2280 Hz

The signal receiver includes a frequency selective circuit that is able to separate out any 2280 Hz component received from the line. The level of this component is then compared with the total signal level. If the receiver detects that the 2280 Hz signal is bigger than the remaining components it then recognizes the presence of a signal and registers the digit being dialled.

Four-tone System

Every time a digit is to be signalled a combination of two tones is transmitted to line. The tones used are given in Table 7.1.

Table 7.1

Digit	Tones		Digit	Tones	
1	697	1209	7	852	1209
2	697	1336	8	852	1336
3	697	1477	9	852	1477
4	770	1209	0	941	1209
5	770	1336	*	941	1336
6	770	1477	#	941	1477

In-band signalling systems suffer from the disadvantage that once a connection has been established further signals cannot be transmitted because the tones would be audible. The use of signalling tones may also cause problems with data transmission over the analogue telephone network.

One advantage of digital transmission using pcm is that all the signalling requirements for 30 channels can be satisfied in one time slot.

Information Theory

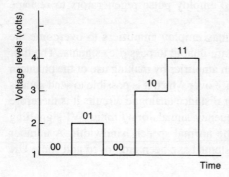

Fig. 7.25

When binary signals are transmitted over a channel at a bit rate of r bits/s the maximum fundamental frequency occurs for alternate 1s and 0s and is equal to $r/2$ Hz. The minimum bandwidth that must be provided is equal to the maximum fundamental frequency, i.e. $B = r/2$ Hz. With this bandwidth r bits/s is the maximum information transmission rate or the *channel capacity*.

If a four-level signal is transmitted in which each level represents a dibit, Fig. 7.25, the capacity is increased to $2r$ bits/s. Doubling the number of signalling levels to eight, so that each level represents a tribit, increases the capacity to $4r$ bits/s and so on.

For an n-level signal the channel capacity will be

$$C = r \log_2 n \quad \text{bits/s} \tag{7.1}$$

or

$$C = 2B \log_2 n \quad \text{bits/s} \tag{7.2}$$

If there was zero noise in a channel it would be possible continually to increase the capacity by employing more and more signalling levels (at the expense, of course, of more complex terminal equipment). In practice, some noise is *always* present in every channel. The effect of this noise on the bit error rate worsens as the distance between adjacent signal levels is reduced.

If the signal power and the noise power at the receiving end of a channel are S and N, respectively, the output signal-to-noise ratio is S/N. The r.m.s. output voltage will be $\sqrt{(S + N)}$ volts, while the r.m.s. noise output voltage will be $\sqrt{N}$ volts. To keep the probability of error to an acceptably low figure, the distance between adjacent signalling levels must be at least $\sqrt{N}$. This means that the maximum number n of possible signalling levels is $n = \sqrt{(S + N)}/\sqrt{N} = \sqrt{1 + S/N}$. Substituting into Equation (7.2)

$$C = 2B \log_2 \sqrt{(1 + S/N)} \quad \text{bits/s} \tag{7.3}$$

or

$$C = B \log_2 (1 + S/N) \quad \text{bits/s} \tag{7.4}$$

EXAMPLE 7.2

Calculate the capacity of a channel of bandwidth 3000 Hz and output signal-to-noise ratio 30 dB.

Solution

30 dB is a power ratio of 1000:1. Therefore

$$C = 3000 \log_2 (1 + 1000)$$

$$= 3000 \frac{\log_{10} 10^3}{\log_{10} 2} = 29\,897 \text{ bits/s} \quad (Ans.)$$

This figure is the theoretical maximum capacity of the channel but to attain it would require the use of $n = \sqrt{(1 + 1000)} \simeq 32$ levels. This is far in excess of any number which is practically possible. If binary signals were to be employed, for example, the maximum bit rate would be only $2B = 6000$ bits/s. Clearly, the capacity of a channel can be increased by increasing its bandwidth and/or its output signal-to-noise ratio. Conversely, for a given value of channel capacity a trade-off between bandwidth and signal-to-noise ratio can always be made.

EXAMPLE 7.3

Calculate the capacity of a channel of bandwidth 1200 Hz and signal-to-noise ratio 20 dB. If the bandwidth is reduced to 1000 Hz what is then the minimum allowable signal-to-noise ratio?

Solution

$$C = 1200 \log_2 (1 + 100) = 7990 \text{ bits/s}$$

When the bandwidth is reduced to 1000 Hz,

$$\frac{7990}{1000} = \frac{\log_{10}(1 + S/N)}{\log_{10} 2}$$

$$\log_{10}(1 + S/N) = 2.4 \text{ and}$$
signal-to-noise ratio $= 250 = 24$ dB $(Ans.)$

8 Digital Modulation

The bandwidth of the commercial-quality speech circuit is 300–3400 Hz but at frequencies above about 3000 Hz group-delay/frequency distortion increases to such an extent that data transmission becomes difficult. Because of this the highest frequency made available for data transmission is usually 3000 Hz. The analogue public switched telephone network (pstn) is unable to transmit signals at or near 0 Hz because of line matching transformers, transmission bridges in the telephone exchange equipment, and the widespread use of amplified circuits, both audio and multi-channel. The direct transmission of a data waveform over a telephone line is only possible for low bit rates up to 150 bits/s over non-amplified lines, or for higher bit rates over *very* short lines. For all other data links the data waveform must be applied to a modem in which a carrier frequency can be modulated in amplitude, frequency, or phase to produce an analogue voice-frequency signal which can be transmitted over the telephone network. As the use of integrated service digital transmission becomes more common the need for modems will decrease.

Modems

A *modem* is a piece of equipment that is fitted to each end of a data circuit to change the serial digital signals produced by the data terminal into voice-frequency signals suitable for transmission over the telephone line network. A modem also converts voice-frequency signals received from line into serial binary data signals which are then passed to the data equipment. A modem may be able to operate in one or more of the *simplex*, *half-duplex* and *full-duplex* modes.

Essentially, a modem has two parts: a transmitter and a receiver. The basic block diagram of the transmitting section of a modem is shown by Fig. 8.1. The encoder is needed when *four-phase*, or *four-level*, modulation is used to group the incoming bits into *dibits* (see p. 95). The carrier frequency is applied to the modulator and is modulated by the data signal to produce the corresponding voice-frequency signal. The bandwidth of the modulated data waveform is

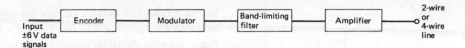

Fig. 8.1 Transmitting section of a modem

limited by the bandpass filter to restrict the transmitted signal to the bandwidth made available by the line. The band-limited signal is amplified before it is transmitted into the two-wire or four-wire line. The amplifier also ensures that the modem is impedance-matched to the line over the range of frequencies to be transmitted.

Figure 8.2 shows the basic block diagram of the receiver section of a modem. The incoming voice-frequency signals are filtered to remove unwanted noise and distortion components lying outside the wanted frequency band. The incoming signals are then amplified by the amplifier whose gain is controlled by an automatic gain control (a.g.c.) circuit to ensure that the voltage input to the demodulator remains more or less constant at all times. The demodulator extracts the information content of the modulated waveform to produce a ±6 V signal; if the output signal is in dibit form it is then decoded into the wanted serial form.

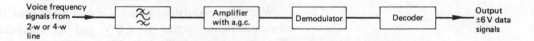

Fig. 8.2 Receiving section of a modem

Not all of the frequency spectrum of a commercial-quality speech circuit can be made available for the transmission of data since it is necessary to avoid those frequencies which are used for signalling and supervisory purposes. For this reason the available frequency spectrum is restricted to 300–500 Hz and 900–2100 Hz, i.e. a bandwidth of 1400 Hz. (2280 Hz is the main signalling frequency used in modern systems.) Low-speed data links of up to 1200 bits/s can be operated so that each bit in the data waveform is represented by a single change in the modulated parameter. Higher-speed modems, operating at 2400 bits/s or more, must generally encode the data waveform before the modulation process is carried out. One method used is to pair bits together to form *dibits*, each dibit being used to produce a single change in the modulated carrier. The four possible dibits are 00, 01, 10 and 11. Since changes in the modulated carrier will now occur only *half* as often as when single bits are used, the baud speed on the transmission medium is reduced by half. The bit grouping process can be carried out still further so that a single change in the modulated carrier occurs for every three bits, or tribits, of information sent; the line baud speed will then be reduced threefold but at the expense of increased circuit complexity and hence increased cost.

The digital modulation methods that are commonly used for data communication are *frequency shift modulation (or keying)*, *differential phase modulation*, *quadrature amplitude modulation* and *vestigial sideband amplitude modulation*.

For bit rates up to and including 1200 bits/s frequency shift modulation is the standard method of modulation used. Frequency shift modulation cannot be used at higher bit rates because the bandwidth needed becomes too great. For bit rates of 2400 bits/s and 4800 bits/s, differential phase modulation is employed. For even higher bit rates, such as 9600 bits/s, 14 400 bits/s or 16 kbits/s, quadrature amplitude modulation is preferred. At the very high bit rates of 48 kilobits/s or more, vestigial sideband amplitude modulation is used.

Frequency Shift Modulation

When a sinusoidal carrier wave is frequency modulated, its frequency is made to vary in accordance with the characteristics of the modulating signal. The amount by which the carrier frequency is deviated from its nominal value is *proportional* to the amplitude of the modulating signal, and the number of times per second the carrier frequency is deviated is *equal* to the modulating frequency.

When the modulating signal is a data waveform, the amplitude of the modulating signal is ± 6 V and its maximum fundamental frequency is one-half of the bit rate. This means that the carrier frequency is *always* deviated to either one of two different frequencies; the *nominal* carrier frequency is the arithmetic mean of these two frequencies but it is never actually present since the modulating signal is never at 0 V. The higher of the two frequencies is used to represent binary 0 while the lower frequency represents binary 1. The system is often known as *fsk*.

The higher the bit rate of the data signal the greater must be the separation between the two frequencies representing binary 1 and binary 0. Otherwise, the detector in the modem receiver will be unable to reliably determine which of the two frequencies is present at any particular instant in time. Because of this and because of the limited bandwidth made available by the pstn the maximum bit rate provided by fsk systems operating over telephone lines is 1200 bits/s.

For lower data rates it is possible to accommodate two channels and thus provide duplex transmission on a two-wire speech circuit. The frequencies used to represent the bits 1 and 0 for the standard bit rates are given by Table 8.1. In each case the nominal carrier frequency is the arithmetic mean of the two frequencies.

An example of a frequency-shift waveform is given by Fig. 8.3. When the data waveform is at $+6$ V (binary 0), the transmitted frequency is the higher of the two frequencies and, similarly, the lower frequency is transmitted to represent binary 1.

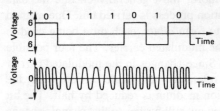

Fig. 8.3 Frequency-shift waveform

Table 8.1

Bit rate (bits/s)	Frequencies (Hz) Binary 0	Binary 1		Nominal carrier frequency
up to 300	1180	980	(different directions of transmission)	1080
	1850	1650		1750
600	1700	1300		1500
1200	2100	1300		1700
75/150 (supervisory channel)	450	390		420

Modulation Index and Deviation Ratio

The modulation index m_f of a frequency-modulated wave and the deviation ratio D of a frequency-modulated wave were defined in Chapter 3 (Equations 3.7 and 3.8).

In the case of frequency shift modulation, the frequency deviation is fixed since the amplitude of the modulating signal voltage — the data waveform — is ±6 V. The maximum modulating frequency exists when the data waveform consists of alternate 1s and 0s and it is then equal to one-half of the bit rate. Thus the deviation ratio of a frequency shift system is given by

$$D = \frac{\text{tone separation}}{\text{bit rate}} \qquad (8.1)$$

EXAMPLE 8.1

Calculate the deviation ratio for a 1200 bits/s frequency-shift data system.

Solution
For a 1200 bits/s data system the two frequencies used are 1300 Hz and 2100 Hz.

Thus, the deviation ratio $= \dfrac{2100 - 1300}{1200} = 0.67$

In this calculation of the deviation ratio the harmonics of the data waveform have not been considered. If this calculation is repeated for the other standard bit rates it will be found that the same answer is obtained.

The Frequency Spectrum of a Frequency-shift Waveform

When a sinusoidal carrier wave of frequency f_c is frequency modulated by a sinusoidal signal of frequency f_m, the modulated wave will contain components at a number of different frequencies. The bandwidth needed for the transmission of a frequency-modulated wave is given by

$$\text{Bandwidth} = 2(f_d + f_m) \tag{8.2}$$

where f_d is the maximum frequency deviation and f_m is the maximum modulating frequency.

For a frequency-shift waveform the bandwidth expression can be rewritten as

$$\text{Bandwidth} = 2(f_d + \text{bit rate}/2) \tag{8.3}$$

$$= \text{tone separation} + \text{bit rate} \tag{8.4}$$

Suppose a 1200 bits/s frequency-shift waveform is to be transmitted. From equation (8.4) the bandwidth needed is

$$\text{Bandwidth} = (2100 - 1300) + 1200$$

$$= 2000 \text{ Hz.}$$

This means that the bandwidth needed to transmit a 1200 bits/s frequency-shift system is greater than the bandwidth that is available (900–2100 Hz or 1200 Hz) when a data system is set up over the pstn. Fortunately, it is not necessary that the waveform of the signal arriving at the far end of a link is undistorted. It is sufficient that the receiving equipment is able to determine at any instant whether a binary 1 or binary 0 bit is being received. This means that it is *not* necessary for all the significant sidefrequencies to be transmitted. Since the deviation ratio of a frequency-shift waveform is approximately unity, most of the energy content of the waveform is concentrated in the carrier and the first-order sidefrequency components. Hence, *only* these components need to be transmitted.

The advantage of only transmitting the first-order sidefrequencies is the considerable reduction in the occupied bandwidth that is obtained. The necessary bandwidth is then only *twice* the maximum modulating frequency or

$$\text{Bandwidth} = \text{Bit rate} \tag{8.5}$$

In the previous example of a 1200 bits/s frequency-shift system the necessary bandwidth is now only 1200 Hz as opposed to the 2000 Hz previously calculated. This narrower bandwidth can be accommodated in the frequency spectrum made available by the pstn (900–2100 Hz) although it should be realized that 1200 bits/s represents the maximum bit rate that can be transmitted.

The basic block diagram of a *frequency shift keying modulator* is shown by Fig. 8.4. The data waveform to be transmitted is band-limited by the input bandpass filter which has a cut-off frequency of 1300 Hz. The waveform applied to the voltage-controlled astable multivibrator consists of the fundamental frequency only of a 1200 bits/s signal or the fundamental plus third harmonic of a

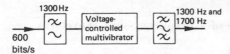

Fig. 8.4 Fsk modulator

600 bits/s signal. The voltage-controlled multivibrator is switched between two states: one in which it oscillates at 1300 Hz and one in which it oscillates at either 1700 Hz or 2100 Hz (depending on the bit rate). The square output waveform of the multivibrator is applied to a bandpass filter whose bandwidth is sufficiently narrow to ensure that only the fundamental frequency of either 1300 Hz or 1700 Hz (or 2100 Hz) is passed. The output of the filter is thus the wanted frequency-shift waveform. Another type of fsk modulator consists essentially of an *LC* oscillator whose tuned circuits values are switched from one set of values to another by the ±6 V data signals.

Phase Shift Modulation

Frequency shift modulation is only available for bit rates up to 1200 bits/s since the next higher standard bit rate of 2400 bits/s would need a bandwidth of 2400 Hz which cannot be provided by the pstn. For 2400 bits/s systems therefore, **phase shift modulation** is used.

When a sinusoidal carrier is phase modulated, its instantaneous phase is made to vary in accordance with the characteristics of the modulating signal. The magnitude of the phase deviation is *proportional* to the modulating signal voltage and the number of times per second the phase is deviated is *equal* to the modulating frequency. In a data system the modulating signal voltage is ±6 V and so the phase deviation obtained is fixed.

Modulating the phase of the carrier will at the same time vary the instantaneous carrier frequency since angular velocity ($\omega = 2\pi \times$ carrier frequency) is the rate of change of phase.

The modulation index of a phase-modulated waveform is, as with frequency modulation, the maximum phase deviation of the carrier frequency produced by the modulating signal. However, the value of the modulation index depends only upon the modulating signal voltage and is quite independent of the modulating frequency.

The modulation process generates a number of sidefrequencies spaced symmetrically either side of the carrier frequency, and the frequency spectrum is exactly the same as that of a frequency-modulated wave having the same numerical value of modulation index.

An example of the use of phase modulation is given by Fig. 8.5. The phase of the carrier is shifted by 180° each time the leading edge of a 0 bit occurs. The phase of the carrier is not altered by a 1 bit. In practice, this form of phase modulation is rarely employed, because detection is difficult, and instead **differential phase modulation** (dpsk) is commonly employed. This version of phase modulation uses *changes* in phase, rather than phase itself, to indicate the dibits 00, 01, 10 and 11.

Four phase changes are used, each of which represents a dibit. The system recommended by the CCITT and now adopted as the standard UK system is given by Table 8.2.

The use of dibits reduces the baud speed on the line and hence the bandwidth that must be made available since the line signal changes only half as often.

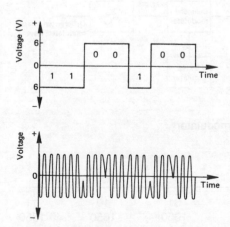

Fig. 8.5 Phase modulated waveform

Table 8.2

Dibit	00	01	10	11
Phase change	+45°	+135°	+315°	+225°

When dibits are used, the equipment at the receiving end of the system must be synchronized with the transmitting equipment for decoding to take place correctly. The necessary synchronization is developed from the changes in phase of the received signal.

Suppose for example that the data signal 10100111 is to be transmitted (lsb first). The bits are grouped together in the decoder section of the modem to form the dibits 11, 01, 10, 10. These dibits are then signalled to line by changing the phase of the carrier by, in turn, +225°, +135°, +315°, +315°. The carrier frequency used is 1800 Hz since at this frequency the group-delay/frequency distortion of a line is small and the frequency is approximately at the middle of the available frequency spectrum.

The basic block diagram of a differential phase modulator is shown in Fig. 8.6. The incoming bit stream is encoded into dibits and then passed on to a digital phase modulator which is fed with a 9 kHz carrier. The output of the modulator is band-limited and then applied to an amplitude-modulation balanced modulator together with the output of a 7.2 kHz oscillator. The balanced modulator produces the sum and the difference of its input frequencies and the 7.2 kHz carrier, and the actual frequencies generated for each dibit are given by Table 8.3.

The bandwidth needed to accommodate the 2400 bits/s dpsk system is 2850 − 750 = 2100 Hz. The required bandwidth is less than would be needed for a frequency-shift system using the same bit rate, but

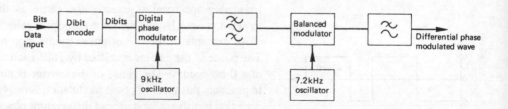

Fig. 8.6 Differential phase modulator

Table 8.3

Dibit	00	01	10	11
Frequencies	750	1050	1650	1350
transmitted (Hz)	1950	2250	2850	2550

it still encroaches into the part of the pstn spectrum occupied by signalling tones. Also, a pstn connection is likely to have a poor group-delay/frequency characteristic at higher frequencies and usually this is much more of a problem than the presence of signalling tones. Many 2400 bits/s systems are therefore operated over privately leased circuits which will probably not have associated in-band signalling equipment and which can be adjusted to have a good group delay characteristic. When a 2400 bits/s system is worked over the pstn, often as a *fall-back* facility for a higher-speed system, it may or may not work satisfactorily depending on the characteristics of the actual link set up. The bandwidth reserved for signalling frequencies is 2130−2430 Hz and, in practice, data signals *can* fall within this band provided there are also signals in the band 900−2130 Hz whose amplitude is not less than a set figure. These will operate guard circuitry in any signalling units that are present and prevent the unit operating.

It is possible to use more than four different phase change values to represent groups of bits and in so doing obtain even greater economy in the use of the available frequency spectrum. If eight phase changes are used, the bits can be grouped in tribits such as 000, 001, etc., and then the effective bit rate for transmission over a line is reduced threefold. This means that a 1200 baud modulation rate can accommodate a 3600 bits/s data waveform, and a 4800 bits/s signal could be sent at a rate of 1600 bauds. Tribit phase changes are given by Table 8.4. Similarly, if the bits are grouped in fours such as 0000, 0001, etc., there are 16 different combinations, and consequently 16 different phase changes are needed. In this case a 1200 baud line will be able to transmit a $4 \times 1200 = 4800$ bits/s signal. Clearly, increasing the number of phases used allows a more efficient usage of the line but this advantage is offset by an increase in the complexity of the receiving equipment. In practice, bit rates in excess of 4800 bits/s generally employ *quadrature amplitude modulation* (qam).

Table 8.4

Tribit	000	001	010	011	100	101	110	111
Phase change	+45°	+90°	+135°	+180°	+225°	+270°	+315°	+360°

Amplitude Shift Modulation

When a carrier wave is amplitude modulated, its amplitude is caused to vary in accordance with the characteristics of the modulating signal (Chapter 3).

A data waveform can be used to switch a carrier wave of appropriate frequency on and off to produce the waveform shown in Fig. 8.7(a). Binary digit 1 is represented by the transmission of the carrier wave

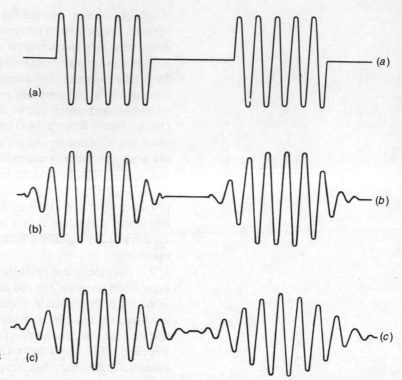

Fig. 8.7 Amplitude-modulated data waveforms

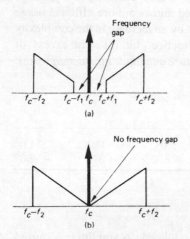

Fig. 8.8 Spectrum diagrams of amplitude-modulated waveforms: (a) when the minimum modulation frequency is f_1; (b) when the minimum modulation frequency is 0 Hz

and binary digit 0 by the carrier being suppressed. Switching the carrier on and off in this way results in the generation of a large number of sidefrequencies and, to restrict the bandwidth occupied, the signal must be passed through a filter. Limiting the number of sidefrequencies transmitted increases the time taken for the envelope of the signal to reach its maximum value. The faster the bit rate, the higher is the fundamental frequency of the data waveform and the more widely spaced are the harmonics contained within that waveform. This means that the higher the bit rate, the more rounded the signal envelope will be; this is shown by Figs. 8.7(b) and (c), in which waveform (c) corresponds to a higher bit rate than waveform (b).

To reduce the required bandwidth still further, **vestigial sideband operation** (vsb) is possible.

In line and radio systems it is possible to suppress the carrier and an unwanted sideband by using the combination of a balanced modulator and a band-pass filter. This practice is possible because the audio bandwidth to be transmitted does not go down to 0 Hz and so a frequency gap exists between the carrier and the edge of the wanted sideband. This is illustrated by Fig. 8.8(a). The fundamental frequency of a data waveform varies according to its information content but may be zero when consecutive 0 or 1 bits are transmitted. This means that the two sidebands produced by the modulation process will not have a frequency gap between them and the carrier (Fig. 8.8(b)).

A filter needs some frequency gap between wanted and unwanted frequencies so that its attenuation can build up and for this reason straightforward ssb operation is not possible for a data system.

In a vsb system, one sideband is transmitted along with a reduced level carrier and a part, or *vestige*, of the other sideband. Vestigial sideband a.m. is used in the UK network for very high speed data links operating at bit rates of 48 kilobits/s and up to 60 kilobits/s. Such high speed links are used to directly connect two digital computers together. The maximum fundamental frequency of a 48 kilobits/s data waveform is 24 kHz, which is far in excess of the bandwidth provided by an audio speech circuit. In any case, in practice frequencies up to 36 kHz are transmitted to line.

If dsb a.m. were to be used, a bandwidth of 72 kHz would be necessary but the use of vsb reduces this to about 44 kHz. A convenient transmission medium which is well able to transmit such a bandwidth is available in the trunk telephone network, namely the line equipment of the 12-channel telephony group. This medium provides a bandwidth of 60–108 kHz. To position the data signal in this bandwidth the signal amplitude modulates a 100 kHz carrier frequency to produce the vsb a.m. waveform shown in Fig. 8.9.

The lower sideband of modulation occupies the band 60–100 kHz and the vestige of the upper sideband that is retained occupies the band 100–104 kHz. This vestige is sufficiently wide to simplify the design of the vsb filters. The carrier frequency is only transmitted at a low level so that it can be used at the receiving end of the system to lock a locally-generated carrier at the correct frequency and phase so it can be used for the demodulation process. A telephone channel can be provided in the band 104–108 kHz but this facility is no longer provided by British Telecom.

Vestigial sideband amplitude modulation is used because of the bandwidth economy it provides; for example, none of the alternative modulation methods previously described could give a 48 kilobits/s system in a 60–108 kHz bandwidth. To greatly reduce the probability of the system causing interference with adjacent 60–108 kHz links carrying multi-channel telephony signals when repetitive data patterns are transmitted, an encoder is fitted at the point at which the data originates to code the signal and break up any repetitive patterns and to spread the power content of the data signal over the available bandwidth. Similar equipment must be fitted at the far end of the data link to decode the incoming signal so as to obtain the original data waveform.

Quadrature amplitude modulation (qam) is a mixture of amplitude and phase modulation used at 9600 bits/s and above. Four phase changes are employed but for each of them there are *two* possible amplitudes. This gives a total of *eight* different states each of which can represent a tribit.

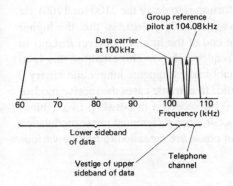

Fig. 8.9 Vsb waveform (from *British Telecom Engineering*)

The Effects of Lines on Modulated Data Signals

When modulated data signals are transmitted over a line, attenuation and group-delay/frequency distortion may distort the received signal to such an extent that the receive modem may not be able to interpret the incoming signals correctly. The error rate will then rise. The effects of the line characteristics will be explained by considering frequency shift modulation.

When frequency shift modulation is used, the higher frequency transmitted will be attenuated to a greater extent than the lower frequency and hence the 0 bits may become of so low a level that it becomes difficult to determine when a 0 bit has been received. This effect is much more prevalent for 1200 bits/s systems than for 600 bits/s systems because the former system uses the higher frequency of 2100 Hz to represent a 0 bit. Thus the effect of line attenuation is to reduce the maximum data transmission rate. The effect of group-delay/frequency distortion is to delay the 2100 (or 1700) Hz pulses to a greater extent than 1300 Hz pulses, so that the higher frequency pulses arrive at the end of the link at incorrect instants in time (relative to the lower frequency bits). The effect of this is to make the pulses overlap so that binary 1 appears longer and binary 0 appears shorter than they should. In extreme cases the receive modem may not be able to detect a 0 bit received in between two 1 bits.

Similar but smaller effects occur when differential phase modulation is used because the changes in phase are signalled to line by various pairs of frequencies.

9 Attenuators, Equalizers and Filters

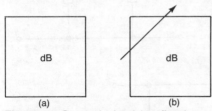

Fig. 9.1 Symbols for (a) a fixed attenuator, (b) a variable attenuator

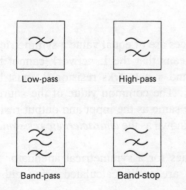

Fig. 9.2 Symbols for types of filter network

Fig. 9.3 Symbols for (a) attenuation equalizer, and (b) group delay equalizer

In both line and radio systems the need often arises for the level of a signal to be reduced, or attenuated, for a group of frequencies contained within a wider band to be transmitted while all other frequencies are suppressed, or for the attenuation/frequency or the group-delay/frequency characteristic of a line to be altered.

An *attenuator* is a circuit that provides a given amount of attenuation to all the frequencies contained in a signal. The symbols for fixed and variable attenuators are given by Fig. 9.1(a) and (b), respectively.

A *filter* is a circuit which has the ability to discriminate between signals at different frequencies because it has an attenuation that varies with frequency in a particular manner. If a signal containing components at a number of different frequencies is applied to the input of a filter, only some of those components will appear at its output terminals, the remainder having been greatly attenuated, and so effectively suppressed.

Four basic types of filter are available for use in telecommunication systems: the low-pass, the high-pass, the band-pass and the band-stop. Figure 9.2 shows the circuit symbols for each of these filters. Filters can be designed using one of the following different techniques: inductor—capacitor filter, crystal filter and active filter.

An *equalizer* is a circuit that is fitted to the end of a line that has the function of making either the overall attenuation/frequency characteristic, or the group-delay/frequency characteristic constant over a given frequency band. The symbol for an attenuation equalizer is shown in Fig. 9.3(a) and the symbol for a group-delay equalizer is given by Fig. 9.3(b).

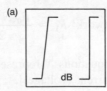
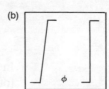

Attenuators

An *attenuator* is a resistive network whose function is to provide a specified amount of loss or attenuation when connected between specified source and load resistances. The input and output terminals of the attenuator are then *matched* to the source and to the load.

Generally, one of three network configurations is employed; these are, respectively, the T, the π, and the L networks. The three networks are shown in Fig. 9.4. Besides providing a specified loss the networks will also match the load to the source.

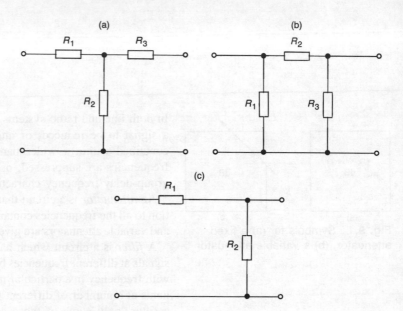

Fig. 9.4 (a) T attenuator, (b) π attenuator and (c) L attenuator

If the source and load resistances are of equal value a symmetrical network must be used. This means that the L network cannot be employed, and for both the T and π networks resistor R_1 must be of the same value as resistor R_3. The common value of the source and the load resistors is then the same as the input and output resistances of the network and this is known as the *characteristic resistance* R_0 of the attenuator.

The necessary resistance values for a symmetrical attenuator to provide a specified attenuation are easily calculated using either equation 9.1 or equation 9.2.

$$\text{T network: } R_1 = \frac{R_0(N - 1)}{N + 1}, \ R_2 = \frac{2R_0N}{N^2 - 1} \tag{9.1}$$

$$\pi \text{ network: } R_1 = \frac{R_0(N^2 - 1)}{2N}, \ R_2 = \frac{R_0(N + 1)}{N - 1} \tag{9.2}$$

In these equations N represents the ratio [input current (voltage)]/[output current (voltage)].

EXAMPLE 9.1

Design a T attenuator to have a loss of 6 dB and a characteristic resistance of 600 Ω.

Solution
$20 \log_{10} N = 6$ dB, hence $N = 2$.
From Equation 9.1,

$$R_1 = \frac{600(2 - 1)}{2 + 1} = 200 \ \Omega \quad (Ans.)$$

$$R_2 = \frac{2 \times 600 \times 2}{4 - 1} = 800 \ \Omega \quad (Ans.)$$

If two, or more, attenuators are connected in cascade their overall attenuation is the arithmetic sum of their individual attenuations, *provided* the attenuators are matched, i.e. have the same values of characteristic resistance.

If an attenuator is not matched to either its source, or its load, or both, the loss obtained will be different from its attenuation by an amount that is not easy to determine. Therefore it is usual to attempt to match attenuators to both their source and their load resistances.

Equalizers

It was seen in Chapter 5 that both the attenuation and the group delay of a transmission line vary with frequency. If the line characteristics are not corrected the signal waveform at the end of the line will be distorted.

An *attenuation equalizer* is a circuit that is fitted at the end of a line, generally just in front of the line amplifier, as shown by Fig. 9.5. The component values of the equalizer are chosen so that the equalizer has a loss/frequency characteristic which is the inverse of that of the line. The overall loss of the line *plus* the equalizer is the *sum* of their individual losses at each frequency, and so the overall loss/frequency characteristic is more or less constant. The process is illustrated by the waveforms given in Fig. 9.5.

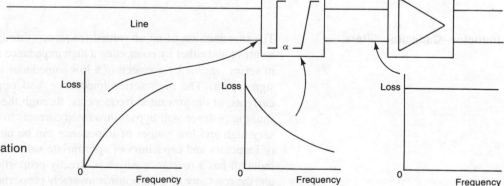

Fig. 9.5 Attenuation equalization of a line

The overall loss of the line is increased by the equalization process, particularly at the lower frequencies, but this is of little account since the equalized signal can easily be given extra amplification. The simplest types of equalizers are employed on audio-frequency lines and Fig. 9.6(a) shows the circuit of one type.

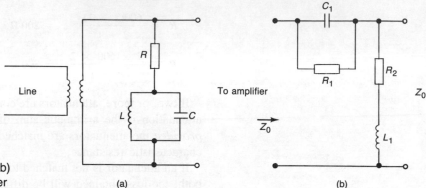

Fig. 9.6 (a) A.F. equalizer (b) Constant-impedance equalizer

At low frequencies the inductor L has little effect and the *basic* loss is provided by the resistor R. With increase in frequency the impedance of the LC parallel circuit increases and so the shunt loss introduced by the equalizer falls. For better results, for music circuits and for multi-channel fdm systems more complex equalizers are employed which are examples of *constant-impedance* equalizers. Figure 9.6(*b*) shows one type that is used in audio-frequency 2-w repeaters. The function of a group-delay equalizer is similar to that of the attenuation equalizer. The equalizer is fitted at the end of the line and it is set to have a group-delay/frequency characteristic which is the inverse of the line's (see Fig. 9.7). Since the group-delay equalizer is required to have a frequency-dependent phase shift and, ideally, no loss, it is constructed using inductors and capacitors only. A wide variety of circuits are possible and Fig. 9.8 gives one possible example.

Inductor–Capacitor Filters

The transmission of an unwanted frequency through a network can be prevented either by connecting a high impedance (at that frequency) in series, and/or by connecting a low impedance in shunt, with the signal path. The high series impedance will oppose the flow of currents, at the unwanted frequencies, through the network, and the shunt impedance will bypass unwanted currents to earth. The necessary high and low values of impedance can be obtained by the use of inductors and capacitors of appropriate value. This is because an inductor has a reactance which is directly proportional to frequency and the reactance of a capacitor is inversely proportional to frequency.

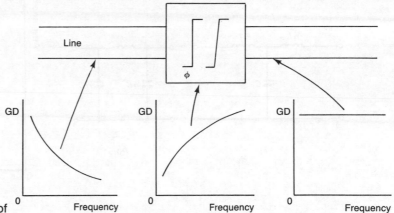

Fig. 9.7 Group-delay equalization of a line

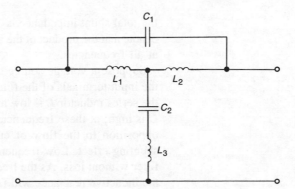

Fig. 9.8 Group-delay equalizer

If a particular band of frequencies is to be transmitted by a filter, series and shunt impedances are required which reach a maximum or a minimum value at the centre of the frequency band. Such impedances can be obtained by the use of series and parallel resonant circuits.

The Prototype or Constant-k Filter

A *low-pass* filter should be able to pass, with zero attenuation, all frequencies from zero up to a certain frequency which is known as the *cut-off frequency* f_c. At frequencies greater than the cut-off frequency the attenuation of the filter will increase with increase in frequency up to a very high value. The basic prototype (or constant-k) T and π low-pass filter circuits are shown in Fig. 9.9(a) and (b) respectively. For both circuits the total series impedance is ωL and

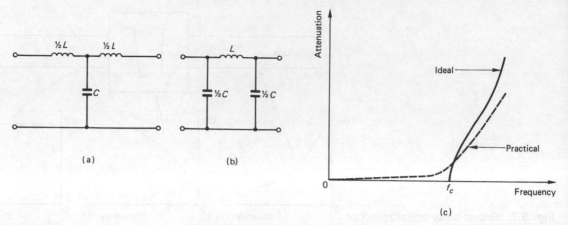

Fig. 9.9 The constant-k low-pass filter

the total shunt impedance is $1/\omega C$. The term 'constant-k' is used to denote that the product of the series and shunt impedances is a constant at all frequencies.

Suppose a voltage source of variable frequency is applied across the input terminals of the filter. At low frequencies the reactance of the series inductor L is low and the reactance of the shunt capacitor C is high; at these frequencies therefore the inductance offers little opposition to the flow of current while the capacitance has zero shunting effect. Low-frequency signals are propagated through the filter without loss. As the frequency of the input signal is increased, the inductive reactance will fall until, at the cut-off frequency f_c, the attenuation of the filter suddenly increases. Thereafter, the attenuation of the filter rises rapidly with increase in frequency. The ideal attenuation/frequency characteristic of a constant-k low-pass filter is shown by Fig. 9.9(c). In practice, an inductor inevitably possesses some resistance and because of this the filter does introduce some attenuation into the passband; also, the attenuation does not rise so sharply at the cut-off frequency. The practical attenuation/frequency characteristic of a low-pass filter is shown by the dotted line of Fig. 9.9(c).

The action of a *high-pass* filter is to transmit all frequencies which are higher than its cut-off frequency and to prevent the passage of all lower frequencies. Figure 9.10(a) and (b) give the circuits of T and π constant-k high-pass filters. At low frequencies the series capacitance C has a high reactance and the shunt inductive reactance is low, so low-frequency signals are attenuated as they travel through the filter. At high frequencies, on the other hand, the series reactance is low and the shunt reactance is high and the filter offers zero attenuation. The attenuation/frequency characteristic of the ideal high-pass filter is shown in Fig. 9.10(c), while the broken curve shows how the presence of resistance modifies the ideal characteristic.

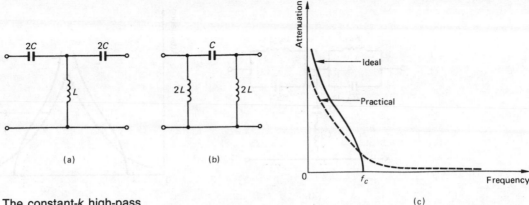

Fig. 9.10 The constant-*k* high-pass filter

Figure 9.11(a) shows the circuit of a T constant-*k* *band-pass* filter and Fig. 9.11(b) shows its ideal and practical attentuation/frequency characteristics. Ideally, the filter passes, with zero attenuation, a particular band of frequencies and offers considerable attenuation to all frequencies outside of this passband. The required characteristic is obtained by using two series-tuned circuits as the series impedance and a single parallel-tuned circuit as the shunt impedance. The three circuits are arranged to be resonant at the same frequency. For signals at or near this common resonant frequency, the series reactance is low and the shunt reactance is high so that the filter offers, ideally, zero attenuation. At frequencies either side of the required passband the tuned circuit's impedances have varied to such an extent that considerable attenuation is offered.

The fourth kind of filter, which is very much less often used, is the *band-stop* filter. This type of filter, shown in Fig. 9.12(a), provides a large attenuation to signals whose frequencies are within a particular frequency band. The ideal and the practical attenuation/frequency characteristics of a band-stop filter are shown in Fig. 9.12(b).

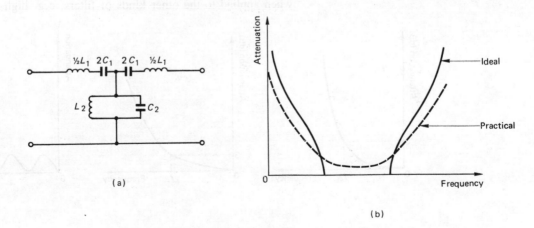

Fig. 9.11 The constant-*k* band-pass filter

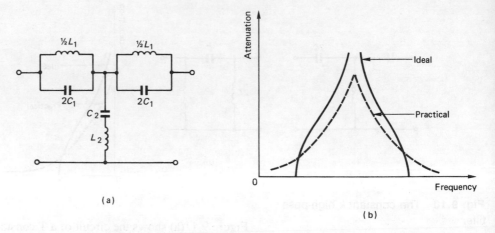

(a)

(b)

Fig. 9.12 The constant-*k* band-stop filter

Modern Filter Designs

In recent years many *LC* filters have been designed using techniques that allow a filter to be designed with a very accurate desired attenuation/frequency characteristic.

The Butterworth attenuation/frequency characteristic of a low-pass filter is shown in Fig. 9.13(a); it is maximally flat in the passband and has an attenuation of 3 dB at the cut-off frequency f_c. The Bessel filter (Fig. 9.13(b)) introduces a constant time delay to all frequencies in the passband and has 3 dB loss at the cut-off frequency. It is evident that the attenuation of the Bessel filter does not rise as rapidly as that of the Butterworth filter. A more rapid increase in attenuation outside of the passband can be obtained by the Tchebyscheff filter (Fig. 9.13(c)) at the expense, however, of ripple in the passband. The relative merits of the three approaches to filter design are the same when applied to the other kinds of filters, e.g. high-pass.

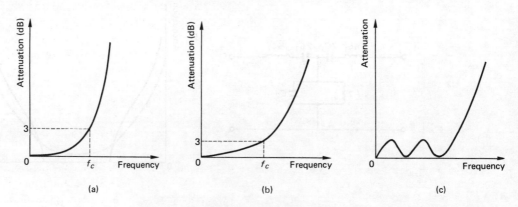

(a)

(b)

(c)

Fig. 9.13 Attenuation—frequency characteristics of (a) Butterworth, (b) Bessel and (c) Tchebyscheff low-pass filters

Crystal Filters

For some applications the maximum selectivity a bandpass *LC* filter can attain is inadequate, and in such cases a crystal filter can be employed. A crystal filter is one in which the required series and shunt impedances are provided by *piezoelectric crystals*.

Piezoelectric Crystals

A piezoelectric crystal is a material, such as quartz, having the property that, if subjected to a mechanical stress, a potential difference is developed across it, and if the stress is reversed a p.d. of opposite polarity is developed. Conversely, the application of a potential difference to a piezoelectric crystal causes the crystal to be stressed in a direction depending on the polarity of the applied voltage.

In its natural state, quartz crystal is of hexagonal cross-section with pointed ends. If a small, thin plate is cut from a crystal the plate will have a particular natural frequency, and if an alterating voltage at its natural frequency is applied across it, the plate will vibrate vigorously. The natural frequency of a crystal plate depends upon its dimensions, the mode of vibration and its original position or *cut* in the crystal. The important characteristics of a particular cut are its natural frequency and its temperature coefficient; one cut, the *GT cut*, has a negligible temperature coefficient over a temperature range from 0 °C to 100 °C; another cut, the *AT cut*, has a temperature coefficient that varies from about +10 p.p.m./°C at 0 °C to 0 p.p.m./°C at 40 °C and about +20 p.p.m./°C at 90 °C. Crystal plates are available with fundamental natural frequencies from 4 kHz up to about 10 MHz or so. For higher frequencies the required plate thickness is very small and the plate is fragile; however, a crystal can be operated at a harmonic of its fundamental frequency and such *overtone* operation raises the possible upper frequency to about 100 MHz.

The electrical equivalent circuit of a crystal is shown in Fig. 9.14. The inductance *L* represents the inertia of the mass of the crystal plate when it is vibrating; the capacitance C_1 represents the reciprocal of the stiffness of the crystal plate; and the resistance *R* represents the frictional losses of the vibrating plate. The capacitance C_2 is the actual capacitance of the crystal (a piezoelectric crystal is an electrical insulator and is mounted between two conducting plates).

A series—parallel circuit, such as the one shown in Fig. 9.14 has two resonant frequencies: the resonant frequency of the series arm RLC_1, and the parallel resonance produced by C_2 and the effective inductance of the series arm at a frequency above its (series) resonant frequency.

If a pair of similar crystals is connected in the series arms of a lattice network and another pair connected in the shunt arms, as shown in Fig. 9.15(a), the network will possess a band-pass characteristic, provided that the series resonant frequency of one pair is equal to the parallel-resonant frequency of the other pair. If the bandwidth

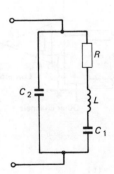

Fig. 9.14 Electrical equivalent circuit of a piezoelectric crystal

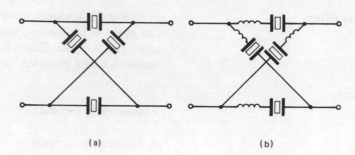

Fig. 9.15 Crystal filters (a) (b)

is not wide enough it can be increased by the connection of an inductor of suitable value in series with the crystal (see Fig. 9.15(*b*)). Crystal filters of the lattice type are widely employed in multi-channel telephony systems. Other, simpler versions of crystal filters are commercially available for use as band-pass filters in radio receivers.

Filters in Parallel

Frequency-division multiplex is the transmission of two or more channels over a single circuit by the positioning of the channels at different parts of the frequency spectrum of that circuit. At the receiving end of an fdm system the received signals must be directed to their correct channels and this is achieved by means of a number of band-pass filters connected in parallel. Since a small frequency gap, about 900 Hz, exists between adjacent passbands, the filters can be paralleled directly and connected to their common load, as shown in Fig. 9.16(a). Each filter is then terminated by the load resistance in parallel with the output impedances of all the other filters. The two filters at the extreme ends of the system bandwidth only have

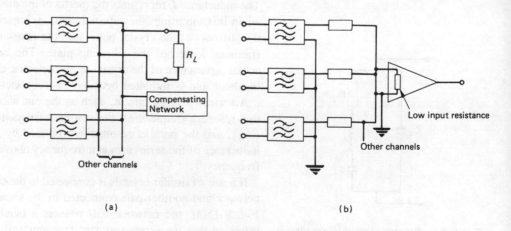

Fig. 9.16 Filters in parallel

another filter connected on one side of them. If these two filters are to be correctly terminated it is necessary to connect a *compensating network* in parallel with the load. This network provides the necessary impedance values at frequencies below the lowest frequency and above the highest frequency passed by the filters, in order for the two extreme filters to be correctly terminated.

In modern equipment a different method of paralleling band-pass filters is employed. The arrangement is illustrated by Fig. 9.16(b). Each filter is terminated by an individual series resistor and the common, low-value input resistance of an amplifier. The value of the series resistor is chosen so that each filter works into its correct load resistance.

Active Filters

Inductors are relatively large and bulky components particularly at the lower frequencies, and also possess core and winding losses that are difficult to predict accurately and which may vary with time, temperature and/or frequency. The need for an inductor in a filter network can be avoided if a resistor—capacitor network is used as the feedback network of an amplifier. A number of different types of active filter are possible but the kind most commonly used since integrated circuit *operational amplifiers* have become readily available is shown in Fig. 9.17. The circuit shown in Fig. 9.17(a) acts as a low-pass filter which can be given a Butterworth, a Bessel or a Tchebyscheff characteristic depending upon the values chosen for the various components. Changing over the positions of the resistors and the capacitors, as in Fig. 9.17(b), produces a high-pass filter with the required type of attenuation/frequency characteristic. Lastly, a band-pass characteristic is obtained by connecting the resistance—capacitance network in the manner shown by Fig. 9.17(c).

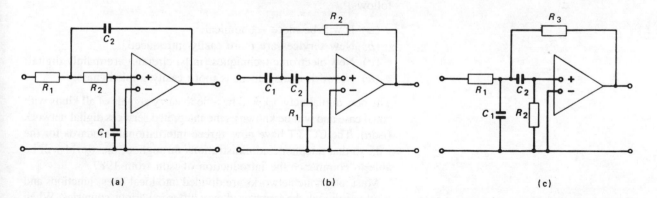

Fig. 9.17 Active filters: (a) low-pass, (b) high-pass and (c) band-pass

10 Two-wire and Four-wire Circuits

The public switched telephone network (pstn) consists essentially of two parts: one part provides switching to organize the required line connections; the other part provides the line network which connects telephone exchanges. Until the 1980s the pstn was entirely based upon analogue switching with a predominantly analogue line network, although pcm systems were being introduced at a fast rate. Digital transmission is now in widespread use throughout the UK and it is anticipated that the trunk network will be entirely digital by the early 1990s. Eventually, the local network will also be converted to digital operation but, because of the sheer size of the operation, this will take some time to complete.

Digital telephone exchanges are replacing analogue exchanges in the UK network at as fast a rate as possible with the intention of eventually converting the pstn wholly to digital operation. The major reasons for the changeover from analogue to digital working are as follows.

(*a*) It will be more economical.
(*b*) New services are more easily introduced.
(*c*) New electronic techniques (in lsi circuits) are mainly digital.
(*d*) Network management is more easily implemented.

The new digital network will be able to carry signals of all kinds with equal ease and will be known as the integrated services digital network (isdn). The CCITT have now agreed international standards for the isdn so that telephone administrations throughout the world will be able to commence the introduction of isdn from 1987.

Most telephone networks are divided into local lines, junctions and trunks, although the terms used may differ in various countries. When analogue techniques are employed, local lines consist of pairs of wires that connect the individual telephone subscribers to their local telephone exchange (le), junctions are two-wire circuits that may or may not be amplified and connect nearby telephone exchanges, and trunks are four-wire circuits that connect more distant exchanges. It

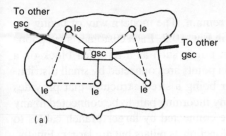

(a)

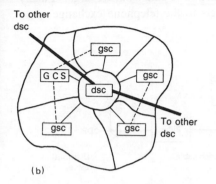

(b)

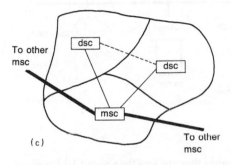

(c)

Fig. 10.1   Trunk switching network

Local Lines

is not economically possible directly to connect every exchange in the network to every other exchange; direct trunks are only provided between two exchanges when justified by the traffic carried. The remainder of the trunk traffic is, as would be expected, between exchanges which are not physically located near one another, and this traffic is routed via trunk switching exchanges known in the UK as *group switching centres* (gsc).

A group switching centre also functions as the local telephone exchange for the area in which it is situated (Fig. 10.1(a)). Each group switching centre collects trunk traffic from the local exchanges in its area and has, in turn, trunks to one or more *district switching centres* (dsc). A district switching centre acts as the trunk switching centre for a number of group switching centres (Fig. 10.1(b)). This stage in the switching network is necessary because it is economically impossible fully to interconnect all the group switching centres.

Lastly, a number of district switching centres are chosen to act as a *master switching centre* (msc) as in Fig. 10.1(c). Each master switching centre acts as a trunk switching centre for a number of district switching centres. All the master switching centres in the network are fully interconnected by direct routes.

Whenever justified economically by the telephone traffic, direct routes are provided between two exchanges, for example, between two group switching centres, between a district switching centre and a group switching centre in another district, or between two district switching centres. These direct routes are known as *auxiliary routes* and they supplement the basic network. It has been estimated that about 85% of trunk traffic is routed over the auxiliary network. Throughout the world, telephone networks are increasingly being converted to digital operation but this chapter will only consider analogue transmission. However, gsc are being replaced by digital principal local exchanges and dsc/msc by digital main switching units.

The connection between a subscriber and his local telephone exchange consists of a pair of wires in a telephone cable. Since a large telephone exchange may have up to 10 000 subscribers the local line network can be quite complicated, particularly because provision must be made for fluctuating demand. The local line network is provided on the basis of forecasts made of the future demand for telephone service, the object being to provide service on demand and as economically as possible. Since the demand fluctuates considerably there is the problem of forecasting requirements and deciding how much plant should be provided initially and how much at future dates. No matter how carefully the forecasting is carried out, some errors always occur and allowance for this must be made in the planning and provision of cable, i.e. the local line network must be flexible. A network must be laid out so that the situation does not arise where potential subscribers cannot be given service in some parts of the exchange area while in other parts spare cable pairs remain. The modern way of laying out a local line network is shown in Fig. 10.2. Each customer's telephone is connected to a distribution point, such as a terminal block on a pole or a wall. The distribution points are connected by small distribution cables to pillars, a pillar being a street structure that provides flexibility because it allows any incoming pair to be connected to any outgoing pair. The pillars are connected by larger branch cables to cabinets; these have the same function as pillars but are larger. Finally, main cables connect the cabinets to the telephone exchange.

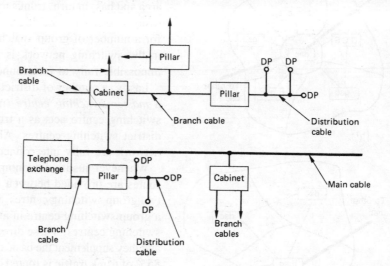

Fig. 10.2 Layout of a telephone exchange area
(DP = distribution point)

Junctions and Trunk Circuits

Junctions interconnect exchanges that are less than about 10 km apart and trunks link exchanges even further apart. Junctions and the shorter trunks are sometimes operated using a two-wire circuit while the longer trunks are all four-wire circuits. Increasingly, pcm systems are employed.

When a signal is transmitted over a telephone line it is attenuated and for all circuits longer than about 30 km will need amplification. Circuits connecting district and/or master switching centres are operated with zero overall loss, group switching centre-to-district switching centre trunks have a loss of 3.5 dB, and group switching centre-to-local exchange junctions have a loss which must not be greater than 4.5 dB. The amplification necessary to achieve these overall loss figures is provided at a number of points along the length of a line.

The required gain cannot be provided simply at either the sending-end or the receiving-end of the circuit. If all the gain were to be provided at the sending-end of the circuit, the signal level would be so high that excessive interference, or *crosstalk*, would be caused to other circuits routed via pairs in the same telephone cable. Conversely, if all of the necessary gain were to be provided at the receiving-end of the line the signal level would have fallen to such an extent that the signal-to-noise ratio (p. 72) would be unacceptably poor. Amplification could not improve the signal-to-noise ratio since both the signal *and* the noise would be equally amplified and the amplifier would itself introduce further noise.

When pcm is used in the analogue network multiplex equipment is necessary at the exchanges, but when pcm is used to provide a junction to a digital exchange the pcm system terminates directly on to a digital switch.

Two-wire Circuits

A two-wire circuit is one that is operated over a single pair of conductors in a telephone cable with signals passing in both directions at the same time and in the same frequency band. When amplifiers of conventional type are employed, the circuit must be split into two separate parts at each amplifier (or repeater) station, each handling one direction of transmission since amplifiers are essentially unidirectional devices.

Figure 10.3 shows a two-wire circuit that has amplification provided at three separate points along its length. At each amplification point the signal path is divided into two by *hybrid transformers* or *terminating units*. A hybrid coil or terminating set (Fig. 10.4) is a device used to convert a two-wire circuit into a four-wire circuit. The operation of the unit is as follows: assume that, as is usually the case, the impedances of the circuits connected across terminals $1-1$ and $2-2$ are equal. A signal applied across the two-wire terminals of the unit will cause a current to flow and this current will induce an e.m.f. into each of the windings connected across terminals $1-1$ and $2-2$. The currents in these circuits are of equal magnitude and so they induce equal e.m.f.s into the balance circuit. The current flowing in the balance impedance is therefore zero. The power contained in the signal applied to the two-wire terminals is divided equally between terminals $1-1$ and $2-2$ so that the loss between these terminals and the input is 3 dB. In addition, transformer losses of about 1 dB are also present.

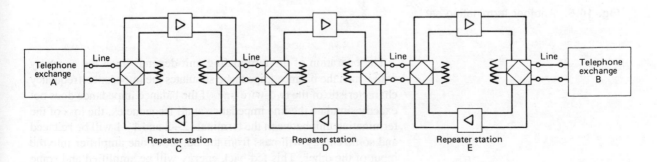

Fig. 10.3 A two-wire amplified circuit

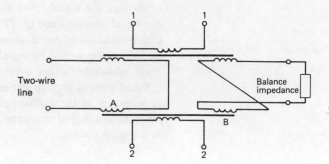

Fig. 10.4 A terminating set

When a signal is applied to the terminals 2—2 of the circuit, the current which flows induces e.m.f.s, with the same polarity, into both winding A and winding B. Currents then flow in the two-wire and balance circuits that induce e.m.f.s of opposite polarity into the winding connected across terminals 1—1. If the balance impedance is adjusted to be equal to the impedance of the two-wire line, these induced e.m.f.s will be equal and they will cancel. Zero current will then flow at terminals 1—1. The signal power applied to terminals 2—2 is divided between the two-wire and balance circuits, giving a total loss of 4 dB between terminals 2—2 and the two-wire terminals. The loss between the terminals 2—2 and 1—1 is very high since little, if any, current flows at terminals 1—1. An alternative circuit for a terminating unit is shown in Fig. 10.5.

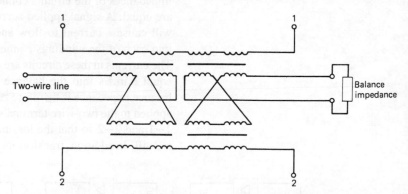

Fig. 10.5 Another terminating set

The operation of the terminating unit depends upon the accuracy with which the balance impedance simulates the impedance/frequency characteristic of the two-wire line. If the balance impedance does not exactly simulate the line impedance at all frequencies, the loss of the terminating unit between the terminals 2—2 and 1—1 will be reduced and some energy will pass from the output of one amplifier into the input of the other. This fed-back energy will be amplified and some of it will appear at the input of the first amplifier and so on. If the circuit is not to oscillate, that is if it is to be *stable*, the loop losses of the circuit must be greater than the loop gains. Since accurate matching between two-wire lines and balance networks at each amplifying point is difficult to achieve in practice, the probability that a two-wire circuit will be unstable increases with increase in the number of amplifiers. It is customary, therefore, to restrict the use of amplified two-wire circuits to lines of such length that amplification at only one point is sufficient.

Amplification may be provided in a two-wire circuit, if the line loss is not greater than 11 dB, by a two-wire repeater inserted as near to the middle of the circuit as possible. The repeater may incorporate conventional amplifiers (Fig. 10.6) or may be a negative-impedance amplifier (Fig. 10.7). The former type consists of two amplifiers and two hybrid coils connected as shown.

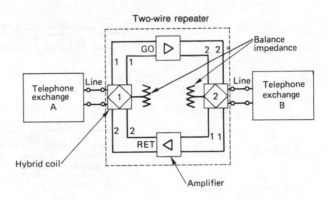

Fig. 10.6 The two-wire amplifier

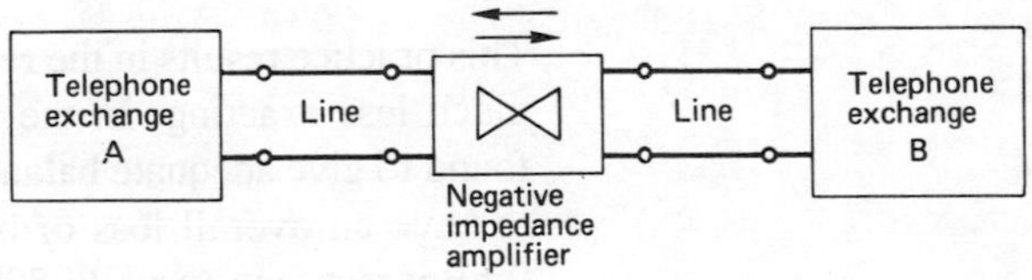

Fig. 10.7 The negative impedance amplifier

Speech signals reaching the repeater from exchange A divide equally at hybrid coil 1, and the signals arriving at terminals 1,1 are 4 dB below the level at the two-wire terminals, and are amplified by the GO amplifier before application to terminals 2,2 of hybrid coil 2. The overall gain of the repeater between the two-wire terminals of the two hybrid coils is equal to the gain of the amplifier minus 8 dB. Typically, the amplifier gain might be 16 dB when an input level of, say, −3 dBm would produce an output level of +5 dBm. The signals appearing at terminals 2,2 of hybrid coil 1 are dissipated in the output impedance of the RET amplifier and serve no useful purpose. Since the loss of a hybrid coil between terminals 2,2 and 1,1 is at least 35 dB, very little energy is fed around the circuit: a necessary condition if the repeater is not to oscillate.

A negative-impedance amplifier (Fig. 10.7) is a circuit that provides amplification of signals in both directions at the same time. Such an amplifier has a gain variable between 2.5 and 12 dB and is designed for use in conjunction with a particular type of audio telephone cable.

All longer analogue trunks are worked four-wire and the basic arrangement of a four-wire circuit is shown in Fig. 10.8. The lines connecting the telephone exchanges to the terminal repeater stations are operated two-wire but at these stations the circuit is split into GO and RETURN paths. Only two terminating units are used, one at each end of the circuit. There is thus only one possible loop path and the possibility of instability is greatly reduced. Also the risk of instability is not increased by increasing the length of the circuit and/or the number of amplifiers. A four-wire circuit is set up to be stable when the two-wire terminals of the two terminating sets are open-circuited.

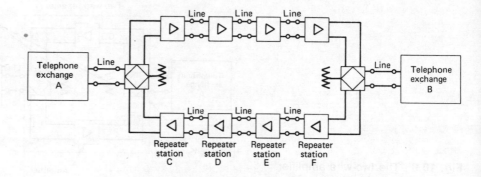

Fig. 10.8 A four-wire amplified circuit

This practice results in the requirements of the two-wire balance being much less exacting. In the majority of cases a 600 ohm resistor is found to give adequate balance. It is usual to set up a four-wire circuit to have an overall loss of between 0 dB and 4.5 dB so that with a 0 dBm test tone (usually 800 Hz) applied to the two-wire terminals of a terminating set, the amplifier output levels, for that direction of transmission, are all +10 dBm. The gain of the amplifiers used is nominally 27 dB and so the minimum input level to an amplifier is −17 dBm.

In practice, the level is very often greater than this and the amplifier must then be preceded by an attenuator of suitable loss. Figure 10.9 shows a four-wire circuit of 3 dB overall loss and gives the levels to be expected at different points in the circuit. The line losses quoted would be at the maximum frequency transmitted, any equalization will be provided before the attenuator preceding the amplifier. A four-wire circuit could become unstable if its loop gain were to increase and become larger than unity. This could happen, for example, if a balance impedance, or two-wire line, were disconnected, and the two-wire/two-wire loss of the circuit was less than 3 dB.

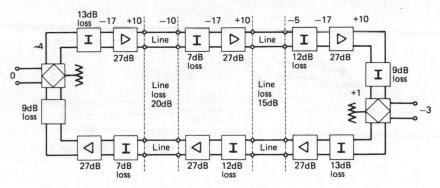

All levels in dBm

Fig. 10.9 Typical losses and gains for a four-wire circuit

Similarly, if either two-wire line became short-circuited at the terminating unit, instability might result. The loop gain of the circuit would also increase if a section of unloaded cable were to be replaced by a length of loaded cable (because of a cable fault) or if the temperature of the cable were reduced. When a circuit becomes unstable it may oscillate; this produces an audible sound that is commonly known as *singing* or *howling*.

So far, the four-wire circuit has been taken as consisting of two pairs in some telephone cables but, in practice, a four-wire circuit may be routed, wholly or partly, over a single channel in a multi-channel telephony system.

A comparison of the relative merits of two-wire and four-wire circuits shows that while the two-wire circuit is economical in the use of cable pairs — an important consideration — the four-wire circuit requires fewer amplifiers for a given length of line and is *much* easier to set up and maintain.

The Interconnection of Junction and Trunk Circuits

The need often arises in a telephone network for trunk and/or junction circuits to be connected in tandem in order to route a call from one exchange to another. Short two-wire circuits are easily switched by manual or automatic means but the problem is somewhat more difficult for four-wire circuits. The automatic equipment required to switch four-wire circuits directly is both complex and costly, and for many years it has been the practice to convert four-wire circuits to two-wire before switching. Figure 10.10 shows a typical connection that involves two local exchanges and two group switching centres. The points marked X denote switching points in telephone exchanges.

The two-wire switching of four-wire trunk circuits suffers from the disadvantages that (*a*) there is a 4 dB loss through each terminating set and (*b*) the overall loss of each trunk circuit cannot be less than about 1.5 dB because of the instability problem and this means that

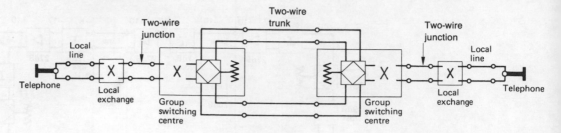

Fig. 10.10 Two-wire switching of trunk circuits

the switching losses cannot be compensated for. These disadvantages naturally assume greater importance as the number of switching points is increased, and therefore all switching in district and master switching centres is carried out on a four-wire basis. The arrangement of a typical four-wire switched connection is shown in Fig. 10.11 in which switching points are again marked X. It can be seen that the connection is four-wire between the two terminal group switching centres and that the group switching centre to local exchange link is worked two-wire. Any losses introduced at the four-wire switching points can be compensated for by increasing the amplifier gains.

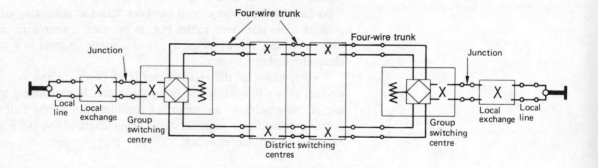

Fig. 10.11 Four-wire switching of trunk circuits

International Circuits

Many international circuits are routed on multi-channel carrier-telephony circuits passing over submarine cable, radio link, or earth satellite and these will be discussed later. Here the intention is to mention the method used to interconnect two subscribers in different countries by a four-wire radio link operating in the high-frequency band 3—30 MHz. Figure 10.12 shows a typical arrangement.

A form of ssb working, known as independent sideband (isb), enables four commercial-quality 25—3000 Hz bandwidth speech circuits to be accommodated in a 12 kHz bandwidth. A subscriber in one country is connected, via the junction and trunk network of

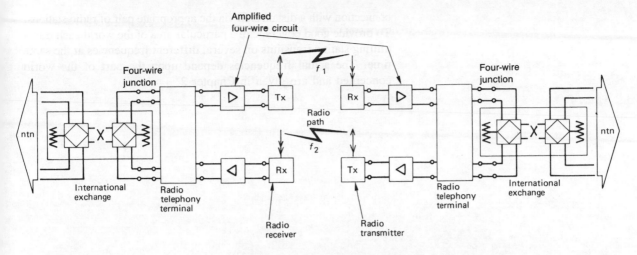

Fig. 10.12 An international telephony circuit
(ntn = national telephone network)

that country, to his international telephone exchange. Here his call is established via a radio circuit to the required country and the telephone network in that country. Signals passing between the two subscribers are at audio frequency up to the radio link itself; at the transmitter the signals amplitude-modulate a carrier in the hf band and the resulting waveform is radiated to the distant receiver. Different frequencies are used for the two directions of transmission over the radio link to eliminate the possibility of singing around the loop.

Ship Radio-telephones

Telephonic communication is often required between a telephone subscriber and a ship at sea and the arrangement for setting up such a connection is shown in Fig. 10.13. The telephone subscriber is connected, via the trunk network, with the control centre. The control centre is linked by cable to a number of coastal transmitting and receiving radio stations, each of which transmits to or receives from a different part of the world. The control centre establishes the required

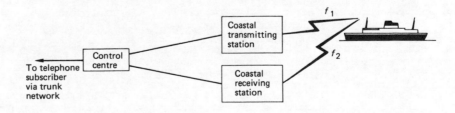

Fig. 10.13 Shore-to-ship telephone connection

connection with a distant ship via the appropriate pair of radio stations. To provide good coverage of a particular area of the world each transmitting station transmits on several different frequencies at the same time. The actual frequencies depend upon the part of the world concerned and are given in Chapter 2.

11 Frequency-division Multiplex Systems

A telephone network connects customers with telephone, telegraph, or data system equipment to any other customer with whom a link is required. When the number of customers is fairly small and they are physically located in the same neighbourhood, it is economically feasible to provide a communication network which consists entirely of physical cable pairs. For example, the telephone subscriber is connected to his local telephone exchange by a pair in the local line network, and can be switched by the exchange equipment to other subscribers connected to that exchange or to a junction to another exchange. The term *space division* is often employed to describe such systems which require a separate cable pair for each circuit, that is systems which do not depend on either frequency-division or time-division. Telephone cable is extremely expensive and, together with the associated ductwork, accounts for the major part of the cost of providing a communication link between two points. It is therefore desirable to increase the traffic-carrying capacity of trunk circuits (in particular) and junctions by the use of either frequency-division multiplex (fdm) or time-division multiplex (tdm) systems. In this chapter the basic principles of multi-channel fdm carrier and coaxial telephony systems will be considered.

Balanced Modulators

Frequency-division multiplex telephony systems utilize single sideband suppressed carrier amplitude modulation of a number of carriers of appropriate frequency. The unwanted sideband is suppressed during the modulation process by means of a *balanced modulator*.

Many balanced modulators, particularly those in multi-channel line systems, do not utilize the square-law characteristics of a diode or transistor but, instead, use the device as an electronic switch. When a transistor or diode is forward biased its resistance is low, and when it is reverse biased its resistance is high. Provided the carrier voltage is considerably greater than the modulating signal voltage, the carrier will control the switching of the device. Ideally, a device should have zero forward resistance and infinite reverse impedance, and this will be assumed in the circuits that follow.

Figure 11.1 shows the circuit of a single-balanced diode modulator.

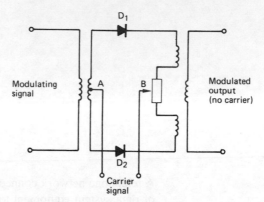

Fig. 11.1 Single balanced modulator

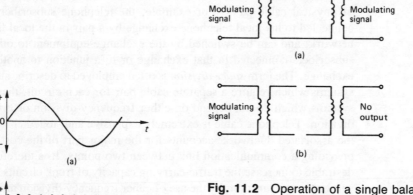

Fig. 11.2 Operation of a single balanced modulator

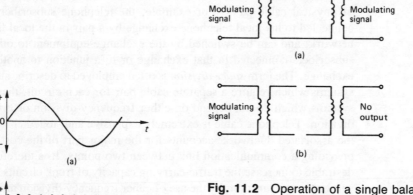

Fig. 11.3 Output waveform of a single balanced modulator
(a) Modulating signal
(b) Carrier wave
(c) Output waveform

During the half-cycles of the carrier waveform that make point A positive with respect to point B, diodes D_1 and D_2 are forward biased and have zero resistance. The modulator may then be redrawn as shown in Fig. 11.2(a); obviously, the modulating signal will appear at the output terminals of the circuit. Similarly, when point B is taken positive relative to point A, the diodes are reverse biased and Fig. 11.2(b) represents the modulator. The action of the modulator is to switch the modulating signal on and off at the output terminals of the circuit. The output waveform of the modulator can be deduced by considering the modulating signal and carrier waveforms at different instants. Consider Fig. 11.3: during the first positive half cycle of the carrier wave a part of the modulating signal appears at the output (1−2 in Fig. 11.3(c)); in the following negative half cycle the modulating signal is cut off (2−3); in the next positive half cycle the corresponding part of the modulating signal again appears at the output terminals (3—4); and so on.

Analysis of the output waveform shows that it contains the upper and lower sidefrequencies of the carrier ($f_c \pm f_m$), the modulating signal f_m, and a number of higher, unwanted frequencies, but the carrier component is *not* present. In practice, of course, diodes are non-ideal and this has the effect of generating further unwanted frequencies and of reducing the amplitude of the wanted sidefrequency. Some carrier leak also occurs, and a potentiometer is often included to adjust for minimum leak.

Another circuit that performs the same function is the *Cowan modulator* (Fig. 11.4). The carrier voltage is applied across points A and B and switches the four diodes rapidly between their conducting and non-conducting states. When point B is positive with respect to point A all four diodes are reverse biased and the modulator may be represented by Fig. 11.5(a); during the alternate carrier half cycles Fig. 11.5(b) applies. The modulator output therefore consists of the modulating signal switched on and off at the carrier frequency. The output waveform is the same as that of the previous circuit (Fig. 11.3) and contains the same frequency components. The Cowan modulator, however, does not require centre-tapped transformers and it is therefore cheaper; it also possesses a self-limiting characteristic (i.e. the sidefrequency output voltage is proportional to the input signal level only up to a certain value and thereafter remains more or less constant).

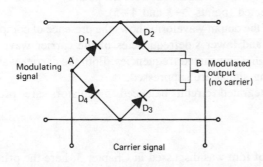

Fig. 11.4 Cowan modulator

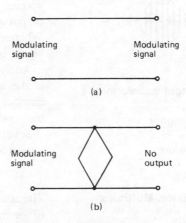

Fig. 11.5 Operation of the Cowan modulator

Sometimes it is necessary to suppress the modulating signal as well as the carrier wave during the modulation process, and then a *double balanced modulator* is used. Fig. 11.6 shows the circuit of a *ring modulator*. During half cycles of the carrier wave when point A is

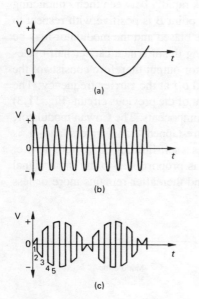

Fig. 11.6 Ring modulator

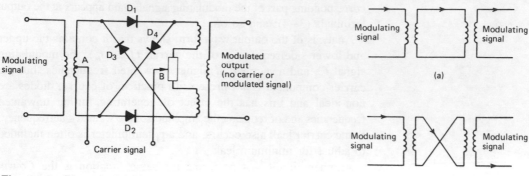

Fig. 11.7 Operation of the ring modulator

positive relative to point B, diodes D_1 and D_2 are conducting and diodes D_3 and D_4 are not; D_1 and D_2 therefore have zero resistance and D_3 and D_4 have infinite resistance; Fig. 11.7(a) applies. Whenever point B is positive with respect to point A, diodes D_1 and D_2 are non-conducting. D_3 and D_4 are conducting, and Fig. 11.7(b) represents the modulator. It is evident that the direction of the modulating signal current at the modulator output terminals is continually reversed at the carrier frequency.

The output waveform of a ring modulator is shown in Fig. 11.8(c) and can be deduced from Figs. 11.8(a) and (b). Whenever the carrier voltage is positive the modulating signal appears at the modulating output with the same polarity as (a) (see points 1−2 and 3−4 at (c)). Whenever the carrier voltage is negative the polarity of the modulating signal is reversed (points 2−3 and 4−5).

Analysis of the output waveform shows the presence of components at the upper and lower sidefrequencies of the carrier wave and a number of higher, unwanted frequencies. Both the carrier wave *and* the modulating signal are suppressed.

Several integrated circuit balanced modulators are presently available.

Fig. 11.8 Output waveform of a ring modulator
(a) Modulating signal
(b) Carrier wave
(c) Output waveform

Frequency-division Multiplex Systems

The concept of fdm was discussed in Chapter 3; here the principles of operation of carrier telephony and coaxial systems will be considered.

The 12-channel CCITT Carrier Group

Most of the coaxial telephony systems in use in Great Britain consist of a suitable combination of a number of CCITT 12-channel carrier groups. Circuits routed over a multi-channel system are operated on

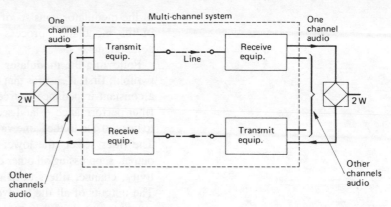

Fig. 11.9 Principle of a multi-channel telephony system

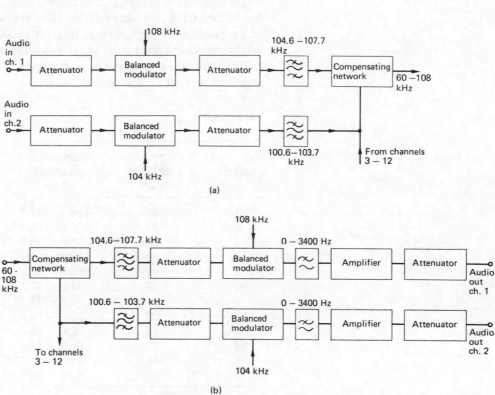

(a)

(b)

Fig. 11.10 The CCITT 12-channel telephony group

a four-wire basis as shown in Fig. 11.9. Figure 11.10(a) is a block schematic of the transmitting equipment required for channels 1 and 2 of the 12-channel group. The audio input signal to a channel is applied to a balanced modulator of the Cowan type together with the carrier frequency appropriate to that channel. The input attenuator ensures that the carrier voltage is 14 dB higher than the modulating signal voltage, as required for correct operation of the modulator. The output of the modulator consists of the upper and lower sideband products of the modulation process together with a number of unwanted components.

Following the modulator is another attenuator whose purpose is twofold; first, it ensures that the following band-pass filter is fed from a constant-impedance source — a necessary condition for optimum filter performance — and second, it enables the channel output level to be adjusted to the same value as that of each of the other channels. The filter selects the lower sideband component of the modulator output, suppressing all other components. To obtain the required selectivity, channel filters utilizing piezoelectric crystals are employed. The outputs of all the twelve channels are combined and fed to the output terminals of the group. The channel carrier frequencies, specified by the CCITT, are listed in Table 11.1. The table also gives details of the passband of each channel filter; it should be noted that the bandwidths correspond to an audio bandwidth of 300−3400 Hz.

The transmitted bandwidth is therefore 60.6−107.7 kHz, or approximately 60−108 kHz.

Table 11.1

Channel no.	Carrier frequency (kHz)	Channel filter passband (kHz)
1	108	104.6−107.7
2	104	100.6−103.7
3	100	96.6− 99.7
4	96	92.6− 95.7
5	92	88.6− 91.7
6	88	84.6− 87.7
7	84	80.6− 83.7
8	80	76.6− 79.7
9	76	72.6− 75.7
10	72	68.6− 71.7
11	68	64.6− 67.7
12	64	60.6− 63.7

The equipment appropriate to channels 1 and 2 at the receiving end of the CCITT 12-channel group is shown in Fig. 11.10(b). The composite signal received from the line, occupying the band 60−108 kHz is applied to the twelve, paralleled, channel filters. Each filter selects the band of frequencies appropriate to its channel, 104.6−107.7 kHz for channel 1, and passes it to the channel demodulator. The attenuator between the filter and the demodulator ensures that the filter works into a load of constant impedance. The demodulator, a Cowan balanced modulator, is supplied with the same carrier frequency as that suppressed in the transmitting equipment. The lower sideband output of the demodulator is the required audio-frequency band of 300–3400 Hz and is selected by the low-pass filter. The audio signal is then amplified and its level adjusted by means of the output attenuator.

The assembly of the basic 12-channel carrier group can be illustrated by means of a frequency spectrum diagram. The spectrum diagram of a single channel is given in Fig. 11.11(a); the actual speech bandwidth provided is 300–3400 Hz but a 0–4000 Hz bandwidth must be allocated per channel to allow a 900 Hz spacing between each channel for filter selectivity to build up. Figure 11.11(b) shows the frequency spectrum diagram for the 12 channels forming a group; the carrier frequency of each channel is given and so are the maximum and the minimum frequencies transmitted. It can be seen that all the channels are *inverted*; that is, the lowest frequency in each channel corresponds to the highest frequency in its associated audio channel, and vice versa. Since all the channels are inverted the group may be represented by a single triangle as shown by Fig. 11.11(c).

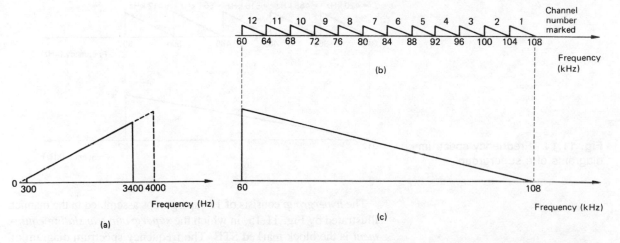

Fig. 11.11 Frequency spectrum diagrams of (a) a commercial-quality speech channel, (b) and (c) a CCITT 12-channel group

The 12-channel system can be more conveniently represented in block diagrams using the system shown in Fig. 11.12. The block marked CTE represents the *channel translating equipment*. The 12-channel group can be used as a building block for the next larger assembly stage or as a system which can be transmitted to line in its own right. This is also true for the larger assemblies that will be described later in this chapter.

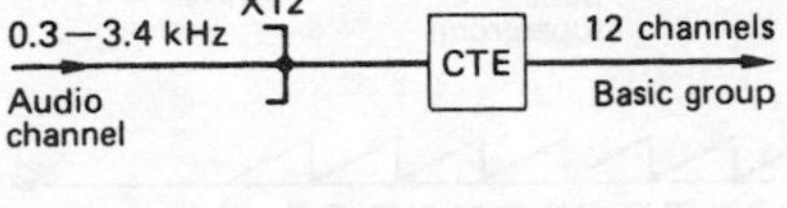

Supergroups and Hypergroups

Five 12-channel groups can be combined to form a 60-channel *supergroup* (Fig. 11.13). The block marked GTE represents the *group*

Fig. 11.12 Representation of the 12-channel group

Fig. 11.13 Formation of a supergroup

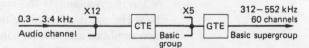

translating equipment. Each of the five groups is used to modulate a different carrier frequency, the particular frequencies being marked on the frequency spectrum diagram of Fig. 11.14(a). Group 1, occupying the frequency band 60−108 kHz, modulates a 420 kHz carrier and the lower sideband of 420−(60−108 kHz) or 312−360 kHz is selected. Group 2 modulates a 468 kHz carrier to produce a lower sideband of 360−408 kHz and so on for the remaining three groups. Since all the groups are erect, the supergroup is erect also as shown by the diagram of Fig. 11.14(b).

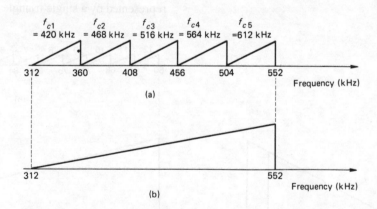

Fig. 11.14 Frequency spectrum diagrams of a supergroup

The *hypergroup* consists of 15 supergroups assembled in the manner illustrated by Fig. 11.15, in which the *supergroup translating equipment* is the block marked STE. The frequency spectrum diagram of a hypergroup is given in Fig. 11.16, and shows the individual supergroups making up the hypergroup. One supergroup is left in the supergroup band of 312−552 kHz and is erect. The other 14 supergroups modulate the appropriate carrier frequencies to position the supergroups in the required parts of the frequency spectrum. These 14 supergroups are all inverted. The bandwidth is nominally 4 MHz.

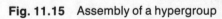

Fig. 11.15 Assembly of a hypergroup

Fig. 11.16 Hypergroup frequency spectrum diagram

A hypergroup can be transmitted to line as a 900-channel system or it may be combined with other hypergroups to produce a system with even larger capacity. For example, a 2700-channel system can be produced by combining three hypergroups in the band 312–12 388 kHz. A modern system combines 12 hypergroups together to assemble a 10 800-channel system occupying the frequency band of 4404–59 580 kHz.

An alternative method of combining supergroups is available and is used in some other countries; it is to assemble five supergroups to form a 300-channel *mastergroup*. The five supergroups modulate, respectively, carrier frequencies of 1364 kHz, 1612 kHz, 1860 kHz, 2108 kHz and 2356 kHz to produce the frequency spectrum diagrams shown in Figs. 11.17(a) and (b). The mastergroup can be transmitted to line or it can be used as the building block for a *supermastergroup*. A supermastergroup is an assembly of three mastergroups and provides 900 channels in the frequency band 8516–12 388 kHz. When required, two or more supermastergroups can be combined to form even larger capacity systems.

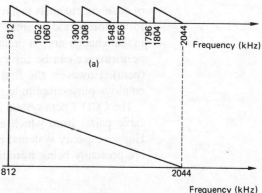

Fig. 11.17 Frequency spectrum diagram of a mastergroup

12 Pulse-code Modulation and Time-division Multiplex

The continuing demand for more and more junction and trunk circuits has led to the widespread use of multiplex telephone systems using either frequency-division or time-division multiplex.

The basic principles of a time-division multiplex (tdm) system were outlined in Chapter 3, where it is shown how tdm allows a number of different channels to have access to the common transmission path for a short period of time. The methods of pulse modulation outlined in that chapter are, in practice, rarely employed since a much better performance can be achieved by the use of *pulse-code modulation* (pcm). However, the first stage in the production of a pcm signal employs pulse-amplitude modulation (pam).

The CCITT pcm system provides 30 channels over audio-frequency cable pairs, from which all the loading coils have been removed. Higher-capacity systems, e.g. 1680 channels, are also available and are presently being introduced into many telephone networks.

Pulse-amplitude Modulation

A time-division multiplex system is based upon the sampling of the amplitude of the information signal at regular intervals, and the subsequent transmission of one or more pulses to represent each sample. In an *analogue* pulse system, for the intelligence contained in the information signal to be transmitted the characteristics of the pulse must, in some way, be varied in accordance with the amplitude of the sample. The pulse characteristic which is varied can be the amplitude, or the width, or the position of the pulse, to give either pulse-amplitude, pulse-duration, or pulse-position modulation. For a *digital* pulse system, i.e. pcm, information about each sample is transmitted to the line by a train of pulses which indicate, using the binary code, the amplitude of the sample. The analogue methods of pulse modulation are rarely used in their own right in modern systems but pulse-amplitude modulation (pam) is employed as a step in the production of a pcm signal.

With pam, pulses of equal width and spacing have their amplitudes

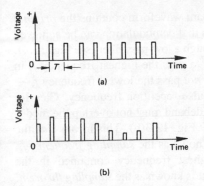

Fig. 12.1 Pulse-amplitude modulation: (a) unmodulated pulses, (b) pam wave

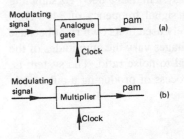

Fig. 12.2 Production of a pam signal: (a) use of an analogue gate, (b) use of a multiplier

varied in accordance with the characteristics of a modulating signal. Figure 12.1(a) shows an unmodulated pulse train, often known as the *clock*, which has a periodic time of T seconds. Thus the number of pulses occurring per second — known as the pulse repetition frequency — is equal to $1/T$. The clock and the modulating signal are applied to the two inputs of an analogue gate, or a multiplier. The gate, or the multiplier, produces an output signal, equal to the instantaneous value of the modulating signal, only when a clock pulse is present (see Fig. 12.2). Thus the pam output of the analogue gate (or the multiplier) consists of successive *samples* of the modulating signal and, assuming a sinusoidal modulating signal, is shown in Fig. 12.1(b).

The pam waveform contains components at a number of different frequencies.

(*a*) The modulating signal frequency.
(*b*) The pulse repetition frequency and upper and lower side-frequencies centred about the pulse repetition frequency.
(*c*) Harmonics of the pulse repetition frequency and upper and lower sidefrequencies centred upon each of these harmonics.
(*d*) A d.c. component whose voltage is equal to the mean value of the pam waveform.

The spectrum diagram of a pam waveform is shown in Fig. 12.3(a). The modulating signal and each of the sidebands are represented by truncated triangles in which the vertical ordinates are made proportional to the modulating frequency, and no account is taken of amplitude. This method of representing signals shows immediately which sideband is erect and which is inverted. The modulating signal occupies the frequency band $f_2 - f_1$ and so the upper and lower sidebands of the pulse repetition frequency are

$$(f_s + f_2) - (f_s + f_1) \quad \text{and} \quad (f_s - f_1) - (f_s - f_2)$$

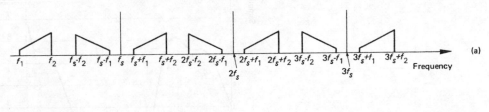

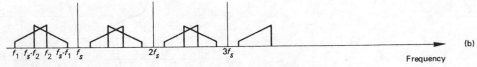

Fig. 12.3 Spectrum diagrams of a pam signal: (a) sampling frequency f_s greater than twice the highest modulating frequency, (b) sampling frequency f_s less than twice the highest modulating frequency

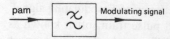

Fig. 12.4 Demodulation of a pam waveform

The frequency spectrum of a pam waveform contains the *original modulating signal* and this means that demodulation can be achieved by passing the pam waveform through a low-pass filter (see Fig. 12.4). The low-pass filter must be able to pass the highest frequency f_2 in the modulating signal but it must *not* pass the lowest frequency $f_s - f_2$ in the lower sideband of the pulse repetition frequency. Clearly, for this to be possible the lower sideband must not overlap the modulating signal in the way shown by Fig. 12.3(b). This means that the pulse repetition frequency, also known as the *sampling frequency f_s*, must be *at least twice* the highest frequency contained in the modulating signal; this requirement is known as the *sampling theorem*. In practice, the sampling frequency must be somewhat higher than twice the highest modulating signal frequency in order to provide a frequency gap in which the filter can build up its attenuation. For example, the CCITT 30-channel pcm system uses a 8000 Hz sampling frequency for a highest modulating signal frequency of 3400 Hz.

Pulse-amplitude modulation is rarely used in its own right because unwanted noise and interference voltages vary the amplitudes of the pulses and degrade the system signal-to-noise ratio. The system is, however, an essential part of the process of producing a pulse code modulated signal.

Pulse-code Modulation

In pulse-code modulation the total audio-frequency amplitude range to be transmitted by the system is divided into a number of allowable voltage levels. Each of these levels is allocated a number as shown by Fig. 12.5 in which eight levels have been drawn. The analogue signal is sampled at regular intervals to produce a pulse-amplitude modulated waveform, but then the pulse amplitudes are rounded off

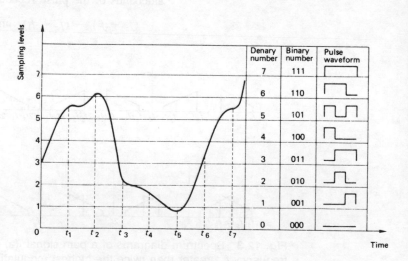

Fig. 12.5 Quantization of a signal

to the nearest allowed voltage level or *quantum*. The process of approximating the sampled signal amplitudes is known as *quantization* and the allowed voltage levels are called quantization levels. The number of the nearest quantization level to each sampled amplitude is encoded into the equivalent binary form and it is then transmitted to the line. Usually, a pulse is sent to represent binary 1 and no pulse is sent to represent binary 0. The pulses are of constant amplitude and width. The advantage gained by the use of constant amplitude pulses is that noise can be eliminated.

Consider Fig. 12.5. The signal waveform is sampled at time intervals t_1, t_2, t_3, etc. At time t_1 the instantaneous signal amplitude is between levels 5 and 6 but, since it is nearer to level 6, it is approximated to this level. At instant t_2 the signal voltage is slightly greater than level 6 but is again rounded off to that level. Similarly, the sample taken at time t_3 is represented by level 2, the t_5 sample by level 1, and so on. Each quantized sample is then *encoded* into the binary-coded pulse waveform shown alongside. The binary pulse train which would represent this signal is shown in Fig. 12.6. A space, equal in time duration to one bit, has been left in between each binary number.

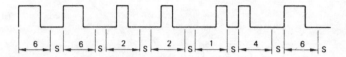

Fig. 12.6 Binary pulse train representing the signal shown in Fig. 12.5

Since the signal information is transmitted in binary form, the number of levels used is always some power of 2^n starting from zero; thus the highest numbered level is $2^n - 1$, where n is the number of bits used to represent each sample. In Fig. 12.5, eight or 2^3 levels only were drawn for clarity but practical systems employ many more quantization levels. For example, the CCITT 30-channel system has 256 levels so that $n = 8$.

Quantization Noise

The signal received at the far end of the system is not an exact replica of the transmitted signal but, instead, is the quantized approximation to it. Fig. 12.7 shows a particular modulating signal and its quantized approximation. The difference between the two is an error waveform which, since it sounds noisy at the output of the system, is known as *quantization noise*. The magnitude of the quantization error depends upon the number of quantization levels used *and* the sampling

Fig. 12.7 Modulating signal and its quantized version (a), (b), and (c) showing the effect of increasing the number of quantization levels; (b) and (d) show the effect of increasing the sampling frequency

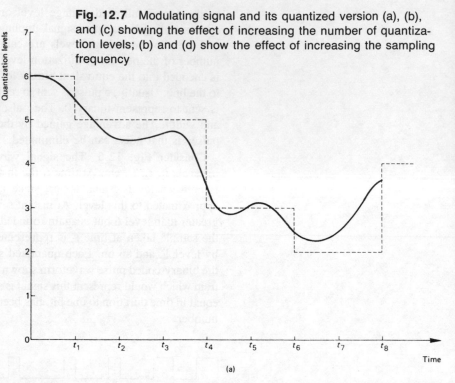

(a)

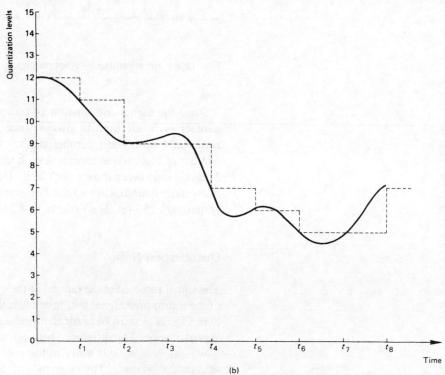

(b)

Fig. 12.7 (cont.)

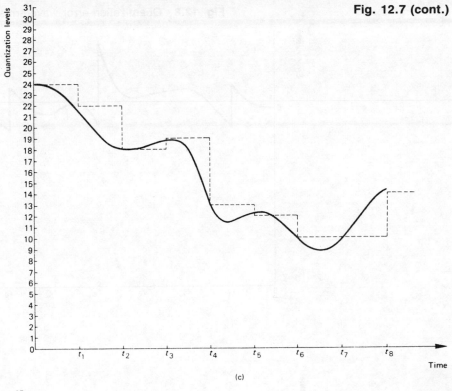

(c)

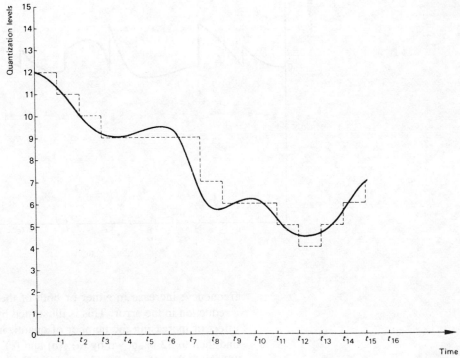

(d)

Fig. 12.8 Quantization error waveforms for Fig. 12.7

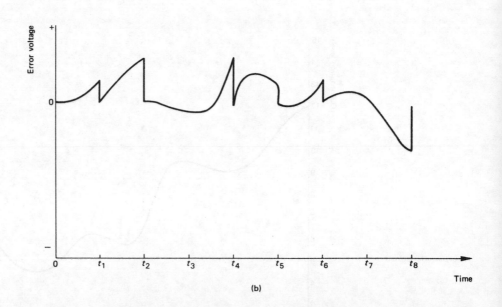

(a)

(b)

frequency; increase in either or both of these parameters produces a reduction in the error. This is illustrated by Fig. 12.7 in which the effect of increasing the number of quantization levels from 8 to 16 and then to 32 is shown by (a), (b) and (c). Also Figs. 12.7(b) and (d) show the effect of an increase in the sampling frequency. The error waveforms for these four cases are given in Fig. 12.8.

Fig. 12.8 (cont.)

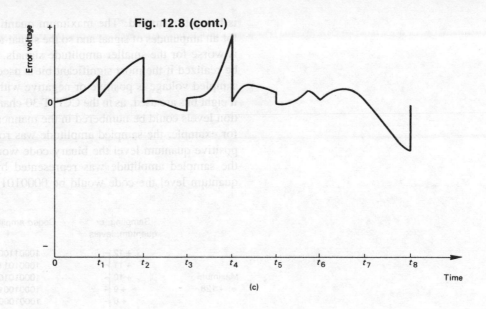

(c)

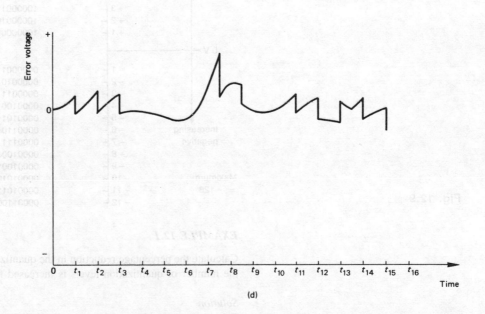

(d)

Unfortunately, as will be shown later, increasing the number of quantization levels and/or increasing the sampling frequency results in a wider bandwidth requirement.

The output waveform of a pcm system can be regarded as consisting of the original modulating signal plus quantization noise. Quantization noise is only present at the output of a pcm system when a signal is being transmitted. The maximum quantization error is the same for all amplitudes of signal and so the signal-to-noise ratio at the output is worse for the smaller amplitude signals. Some improvement can be realized if the most significant bit is used to indicate whether the sampled voltage is positive or negative with respect to earth. Thus, if eight bits are used, as in the CCITT 30-channel system, the quantization levels could be numbered in the manner shown in Fig. 12.9. If, for example, the sampled amplitude was rounded off to the twelfth positive quantum level the binary code would be 10001100, and if the sampled amplitude was represented by the eleventh negative quantum level the code would be 00001011.

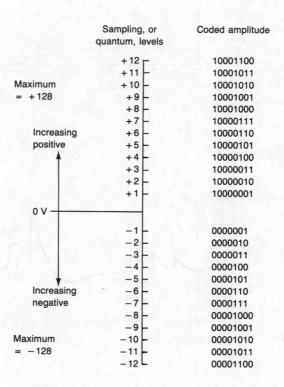

Fig. 12.9

Sampling, or quantum, levels	Coded amplitude
+12	10001100
+11	10001011
+10	10001010
+9	10001001
+8	10001000
+7	10000111
+6	10000110
+5	10000101
+4	10000100
+3	10000011
+2	10000010
+1	10000001
−1	0000001
−2	0000010
−3	0000011
−4	0000100
−5	0000101
−6	0000110
−7	0000111
−8	00001000
−9	00001001
−10	00001010
−11	00001011
−12	00001100

Maximum = +128. Increasing positive. 0 V. Increasing negative. Maximum = −128.

EXAMPLE 12.1

Calculate the percentage reduction in the quantization error, or noise, when the number of quantization levels is increased from 256 to 512.

Solution
The increase in the number of sampling levels is twofold. Hence the quantization error is reduced by one half or 50%. (*Ans.*)

Non-linear Quantization

The use of equally spaced quantization levels as so far assumed is useful as long as the signal amplitude is large enough to cover several quantization levels. For small-amplitude speech signals, however, the output signal-to-noise ratio may well be inadequate. The signal-to-noise ratio for small signals can be improved, without increasing the number of quantization levels, by the use of non-linear quantization. The quantization levels are no longer equally spaced but, instead, are much closer together near the middle of the amplitude range than near the two ends of the range. This will ensure that the small-amplitude signals are precisely quantized, while the larger-amplitude signals are quantized more coarsely. The spacing of the quantization steps is arranged to follow a logarithmic law such that a more or less constant output signal-to-noise ratio is obtained for signals of all allowable amplitudes.

The improvement in the quantization accuracy that non-linear quantization can give is illustrated by Fig. 12.10. The same small-voltage signal has been applied to a linear quantizer in Fig. 12.10(a) and to a non-linear quantizer in Fig. 12.10(b). It is clear that the quantization error is much smaller in the second case.

Non-linear quantization can be obtained in two different ways:

(1) the analogue signal can have its range of amplitudes compressed and then encoding can take place, or
(2) a non-linear encoder can be employed.

Time-division Multiplex

The general principles of time-division multiplex have been described earlier in this book. Here tdm will be discussed in terms of the CCITT 30-channel pcm system.

The minimum sampling frequency, i.e. the number of samples taken per second, is equal to twice the highest frequency contained in the analogue signal. Thus, if the highest modulating signal is limited to 3400 Hz by an input low-pass filter, the minimum sampling frequency will be 6800 Hz. In practice, a sampling frequency somewhat greater than the minimum allowable would be used. The CCITT system uses a 8000 Hz sampling frequency for the maximum audio frequency of 3400 Hz.

Each sample of the information signal is represented by eight bits, the first bit indicating the polarity of the sampled amplitude. Each bit is 488 ns wide and, hence, each sample occupies a time period of $8 \times 0.488 = 3.9 \, \mu\text{s}$. A sample is taken every 1/8000 s or every 125 μs, and this means that the larger part of each sample period is unoccupied, see Fig. 12.11. The unused time periods can be used to carry other pcm channels; the number which can be fitted into the time available is $125/3.9 = 32$ channels. Two of these channels are used for synchronization and signalling purposes (that is channels 0 and 16) and so 30 channels are available to carry speech signals.

The *basic* block diagram of the system is shown by Fig. 12.12.

The channels are interleaved sequentially, sample by sample, and the time period occupied by 32 time slots is known as a *frame*. Thus

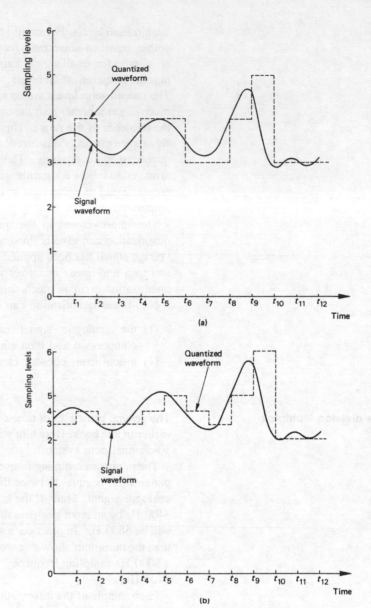

Fig. 12.10 Showing the improvement in quantization accuracy obtained by the use of non-linear quantization: (a) linear, (b) non-linear quantization

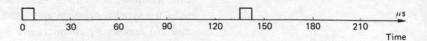

Fig. 12.11 Showing the time that a single channel pcm system occupies a transmission path

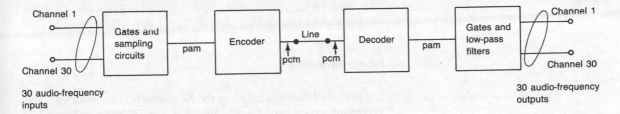

Fig. 12.12 Basic block diagram of 30-channel pcm system

a frame occupies 125 μs and contains $32 \times 8 = 256$ bits. The line bit rate is

$$256 \times 8000 \quad \text{or} \quad 2048 \text{ kilobits/s (2.048 Mb/s)}$$

Bandwidth of a Pulse-code Modulated Waveform

The allowable amplitude range of a pulse-code modulation system is divided into 2^n quantization levels. Each time the analogue signal is sampled, n bits are transmitted to indicate the appropriate quantum level. The analogue signal is sampled f_s times per second, where f_s is the sampling frequency, and so the number of bits transmitted per sample per channel is nf_s.

If the tdm system has m channels then the bit rate is

$$\text{Bit rate} = mnf_s \qquad (12.1)$$

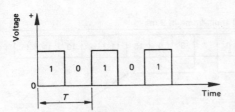

Fig. 12.13 Pcm waveform having the highest fundamental frequency

i.e. bit rate = number of channels × sampling frequency × number of bits per word. The maximum fundamental frequency of the transmitted binary pulse train occurs when it consists of alternate 1s and 0s as shown in Fig. 12.13. The maximum fundamental frequency is the reciprocal of the periodic time of the pcm waveform and is equal to one-half of the bit rate. The minimum frequency of the pcm waveform occurs when the binary number consists of consecutive 1 or 0 bits and is equal to 0 Hz. The minimum bandwidth which must be provided for a pcm waveform, assuming that only the fundamental frequency component need be transmitted and the system is noise free, is thus from 0 Hz to (bit rate)/2 Hz.

$$\text{Minimum bandwidth} = \frac{\text{bit rate}}{2} \qquad (12.2)$$

EXAMPLE 12.2

Calculate (*a*) the bit rate per channel, (*b*) the total bit rate, and (*c*) the minimum bandwidth needed for the CCITT 30-channel pcm system.

Solution
In this system 256 quantum levels are indicated by eight bits. Hence, since the sampling frequency is 8 kHz

(*a*) bit rate per channel $= 8 \times 8000 = 64$ kilobits/s   (*Ans.*)

(*b*) total bit rate   = 64 kilobits/s × 32 = 2.048 Mb/s   (*Ans.*)
(*c*) minimum bandwidth = bit rate/2 = 1.024 MHz   (*Ans.*)

Frame Structure

The period of time occupied by the 32 channels is known as a *frame*. The frame structure of the 30-channel system is shown in Fig. 12.14. Sixteen frames form one *multi-frame* which occupies a time period of 2 ms. Frame 0 has its first four bits used for multi-frame alignment and some, or all, of its remaining bits used for alarm and supervisory purposes. Frames 1 to 15 are the signal frames and an exploded view of frame 7 is given. It can be seen to contain 32 time slots (0−31) each of which occupies a time period of 3.9 μs. Time slot 0 contains the frame synchronization signals and time slot 16 is (mainly) devoted to carrying telephone exchange signalling information. For frame 0, only the time slot 16 carries a multi-frame synchronization signal. Thus, 30 channels are made available for the transmission of speech. Synchronization is necessary to ensure that the information transmitted by, say, channel 1, is directed to channel 1 at the receiving end of the system.

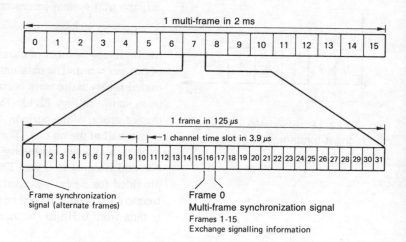

Fig. 12.14 30-channel pcm frame structure

Signalling

The 64 kb/s capacity of the signalling channel means that it is able to satisfy all of the signalling requirements of the 30 speech channels with *no* interference with any speech or data transmissions.

Each speech channel is allocated four signalling bits so that 120 bits are needed. This demands that 16 frames, giving 16 × 8 = 128 bits, are used. The first and last frames (0 and 15) carry frame synchronization bits; the other frames each carry the signalling bits of two adjacent channels. Thus frame 1 serves channels 0 and 16, frame 2 serves channels 2 and 17, and so on.

One separate-channel signalling system uses the four signalling bits for the purposes shown in Table 12.1.

Table 12.1

	Meaning	
Signal bits	Forward direction	Backward direction
0001	trunk offer	manual hold
0011	circuit seized	called number answer
0111	—	circuit free
1001	—	coin fee check
1011	dial break	—
1111	circuit idle	circuit busy

Line Signals

The information about the sampled amplitudes is signalled to the line using the binary code. So far it has been assumed that unipolar pulses are transmitted, i.e. a positive voltage representing binary 1 and zero voltage representing binary 0. This line code has the disadvantages that (a) it always has a d.c. component, (b) it contains low-frequency components of relatively large amplitude. The d.c. and low-frequency components are not wanted because their absence simplifies the design of the pulse regenerators which are fitted along the length of the line.

The disadvantages of unipolar signalling can be overcome by the use of *alternate mark inversion* (ami), and the principle of this method of signalling is shown by Fig. 12.15. The basic unipolar signal (Fig. 12.15(a)) has its alternate bits inverted as shown by (b) and then (Fig. 12.15(c)) a bipolar signal is generated by inverting alternate *marks* or 1 bits. This procedure ensures that each 1 bit is of opposite sign to the preceding 1 bit. Alternate mark inversion reduces the probability of a long stream of 0s being transmitted to line when a channel is idle.

Alternate mark inversion signalling has the disadvantage that a long string of 0s in a bit stream results in no pulses being sent to line. The line pulse regenerators derive their timing information from the pulses incoming to them and so their operation is adversely affected by a period of no pulses. To overcome this problem a mark or 1 bit can be inserted into the bit stream after a given number of 0s have been sent. This kind of signalling is known as *high-density bipolar n* (HDBn). In the 30-channel system, n is chosen to be 3, the encoding being known as HDB3. Thus 3 is the maximum number of consecutive 0s allowed in the transmitted signal.

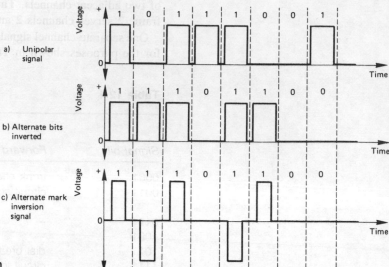

Fig. 12.15 Alternate mark inversion

Pulse Regenerators

In its passage along a telephone line, the pcm signal is both attenuated and distorted but, provided the receiving equipment is able to determine whether a pulse is present or absent at any particular instant, no errors are introduced. To keep the pulse waveform within the accuracy required pulse regenerators are fitted at intervals along the length of the line. The function of a **pulse regenerator** is to check the incoming pulse train at accurately timed intervals for the presence or absence of a pulse. Each time a pulse is detected, a new undistorted pulse is transmitted to line and, each time no pulse is detected, a pulse is not sent.

The simplified block diagram of a pulse regenerator is shown in Fig. 12.16. The incoming bit stream is first equalized and then amplified to reduce the effects of line attenuation and group-delay/frequency distortion. The amplified signal is applied to a timing circuit which generates the required timing pulses. These timing pulses are applied to one of the inputs of two two-input AND gates, the phase-split amplified signal being applied to the other input terminals of the two gates. Whenever a timing pulse *and* a peak, positive or negative, of the incoming signal waveform occur at the same time, an output pulse is produced by the appropriate pulse generator. It is arranged that an output pulse will not occur unless the peak signal voltage is greater than some predetermined value in order to prevent false operation by noise peaks.

Provided the bit stream pulses are regenerated before the signal-to-noise ratio on the line has fallen to 21 dB, the effect of line noise on the error rate is extremely small. This means that impulse noise can be ignored and white noise (i.e. noise of constant voltage over the operating bandwidth) is *not* cumulative along the length of the

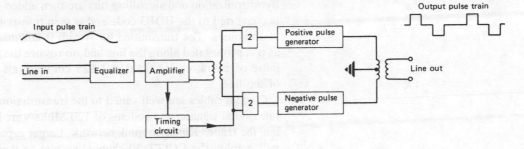

Fig. 12.16 Pulse regenerator

system. This feature is in marked contrast with an analogue system in which the signal-to-noise ratio must always progressively worsen towards the end of the system. Thus, the use of pulse regenerators allows very nearly distortion-free and noise-free transmission, regardless of the route taken by the circuit or its length. The main difficulty that arises is avoiding *jitter*. Jitter is caused by changes in phase in the regenerator which are mainly the result of noise and/or inadequate timing.

Pulse-code Modulation Systems

The basic block diagram of a 30-channel pcm system is shown by Fig. 12.17, in which 30 band-limited (to 4 kHz) channels are sampled sequentially at a sampling frequency of 8000 Hz to produce pam waveforms. The pam waveforms are interleaved on the common highway to produce a time division multiplex signal before the multiplex signal is encoded by the non-linear encoder to give a unipolar pcm signal.

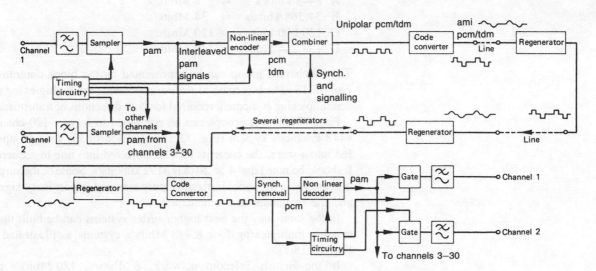

Fig. 12.17 30-channel pcm system

Synchronization and signalling bits are then added before the signal is converted to the HDB3 code and is then transmitted to the line at 2048 kbits/s. The transmitted bit stream is attenuated and distorted as it is propagated along the line and, to restore the pulse waveform, pulse regenerators are fitted at 1.828 km intervals along the length of the line.

Coaxial cables are well suited to the transmission of very high bit rate digital signals, and systems of 120 Mbits/s are being introduced into the United Kingdom trunk network. Larger capacity pcm systems will employ the CCITT 30-channel system as their basic building block.

The old 24-channel pcm system is very similar to the 30-channel system, the main differences being that, in 24-channel,

(a) a compressor followed by a linear encoder is used to get the required non-linear encoding;
(b) the transmitted bit stream uses the alternate mark inversion code;
(c) the line bit rate is 1536 kbits/s.

Higher-order TDM Systems

The CCITT 30-channel pcm system can be used as the basic building block for higher-order tdm systems in a similar manner to the way in which large-capacity fdm systems are built up from the 12-channel group.

A 30-channel frame contains $32 \times 8 = 256$ bits and so the line bit rate is

$$256 \times 8000 = 2048 \text{ kbits/s} \equiv 2 \text{ Mbits/s}$$

The higher-order bit rates recommended by the CCITT are

A 8448 kbits/s $\equiv$ 8 Mbits/s
B 34 368 kbits/s $\equiv$ 34 Mbits/s
C 120 000 kbits/s $\equiv$ 120 Mbits/s
D 139 264 kbits/s $\equiv$ 140 Mbits/s

The 30-channel group can be represented by the block diagram of Fig. 12.18. The box marked *muldex* represents the multiplexing and demultiplexing equipment required for both directions of transmission.

Four 30-channel groups can be combined to form a 120-channel 8448 kilobits/s system (Fig. 12.18(b)). Each of the four groups is fed into a store, the contents of which are fed into line in sequence. It should be noted that 4×2048 is 8192 kilobits/s. Some of the surplus bits are used for synchronization purposes and supervisory signals and the remainder are redundant.

In the same way the next higher-order systems can be built up by suitably multiplexing four 8.448 Mbits/s systems as illustrated by Fig. 12.18(c).

In the British Telecom network, 8 Mbits/s, 120 Mbits/s and 140 Mbits/s systems have been installed.

Fig. 12.18 Assembly of higher-order pcm systems: (a) 30-channel; (b) 120-channel; (c) 1960-channel

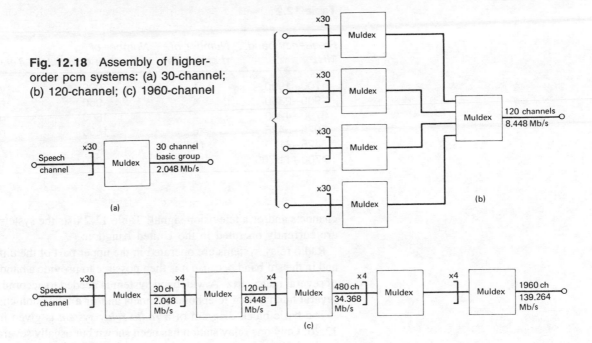

(a)

(b)

(c)

Relative Merits of FDM and TDM

The advantages of tdm, assuming the use of pcm, are as follows.

(*a*) The use of regenerators allows an almost distortion-free path regardless of distance, whereas in an fdm system the signal-to-noise ratio gets progressively worse with increase in distance.

(*b*) Channel selection is achieved by the use of relatively cheap electronic gates and low-pass filters rather than with expensive crystal filters.

(*c*) The level and phase of the received signal do not depend upon the stability of gain and/or phase shift of the circuits it has been transmitted through.

(*d*) In an fdm system, care must be taken to avoid non-linearity in amplifiers, etc., because this will result in intermodulation and crosstalk between channels. This is not so for a tdm pcm system.

The main disadvantage of pcm tdm is that a greater bandwidth is needed to transmit a given number of channels than required for the corresponding fdm system. If the signal-to-noise ratio on a pcm transmission path falls below 21 dB, the received signal will deteriorate rapidly and become worse than the corresponding fdm signal.

Radio Relay Systems

Radio relay systems using line-of-sight transmissions in the uhf and shf bands (Fig. 12.19) can provide a large number of telephone

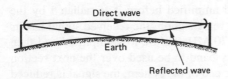

Fig. 12.19 The space wave

Table 12.2

Frequency band MHz	Number of rf channels	Number of telephone channels	Television
1700–1900	1	960	yes
1900–2300	4	960	no
3790–4200	4	1800	no
5925–6425	6	1800	no
6425–7110	12	960	yes
10 700–11 700	10	960	yes

channels and/or a television signal. Table 12.2 lists the systems that are currently operated in the United Kingdom.

Radio relay systems are operated in the upper part of the uhf band and in the shf band because it is then possible to provide a bandwidth of several megahertz. A wideband system is needed to accommodate several hundreds of telephone channels and/or a television channel.

The basic block diagram of a *radio relay system* is given in Fig. 12.20. Only one relay station has been shown but usually several will be used, the actual number depending upon the length of the route.

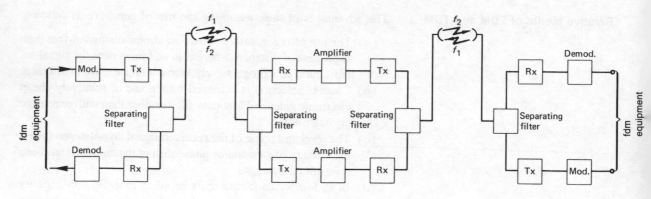

Fig. 12.20 Radio relay system (Rx = receiver, Tx = transmitter)

At the transit terminal, the baseband signal (the signal produced by a tdm or an fdm multi-channel system or by a television camera) is pre-emphasized and is then used to frequency-modulate a 70 MHz carrier. The modulated wave is then shifted to the allocated part of the frequency spectrum and amplified before it is radiated by the parabolic dish aerial. At the relay station, the received signal has its frequency changed back to 70 MHz before it is amplified and then shifted back to the frequency band to be used over the next section of the route. At the receiving end of the system, the signal is reduced in frequency to 70 MHz before it is demodulated.

Both radio relay and line systems are widely used as integral parts of the national telephone network. The two systems have a number of advantages and disadvantages relative to one another which often means that one or the other is best suited for providing communication over a given route. The relative merits of the two systems are listed below.

(a) A radio relay system is generally quicker and easier to provide.

(b) The problems posed by difficult terrain are easier to overcome using a radio relay system.

(c) It is easier to extend the channel capacity of a radio relay system.

(d) Difficulties may be experienced in obtaining suitable (line-of-sight distance) sites for a radio relay system.

(e) When relay station sites have been chosen it may be difficult to gain access to them, whereas coaxial systems usually follow roads and so access is relatively easy.

(f) The transmission performance of a radio relay system is adversely affected by bad weather conditions.

Communication Satellite Systems

Most of the long-distance international telephone traffic which is not carried by submarine cable systems is routed via a broadband communication satellite system, the basic principle of which is illustrated by Fig. 12.21. The ground stations are fully integrated with their national telephone networks and, in addition, the European ground stations are fully interconnected with one another. Four frequencies are used; the North American ground station transmits on frequency f_1 and receives a frequency f_4, the European stations transmit frequency f_3 and receive frequency f_2. Essentially, the

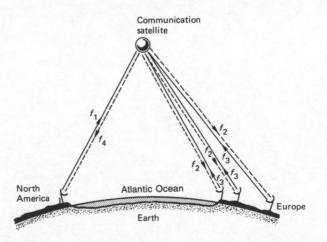

Fig. 12.21 Communication satellite system

purpose of the communication satellite is to receive the signals transmitted to it, frequency translate them to a different frequency band (f_1 to f_2 or f_3 to f_4), amplify the signals, and then re-transmit them to the ground station at the other end of the link.

Communication satellites which form an integral part of the public international telephone network are operated on a global basis by COMSAT (Communication Satellite Corporation) on behalf of an international body known as INTELSAT (International Telecommunication Satellite Consortium). The COMSAT system employs communication satellites travelling in the circular equatorial orbit at a height of 35 880 km. This particular orbit is known as the synchronous orbit because a satellite travelling in it appears to be stationary above a particular part of the Earth's surface. Seven satellites are used, positioned around the Earth so that nearly all parts of the surface of the Earth are 'visible' from at least one satellite. A large number of ground stations are in use and now number more than a hundred in nearly a hundred different countries.

Each ground station transmits its telephone traffic to a satellite on the particular carrier frequency allocated to it in the frequency band 5.925−6.425 GHz. This is a bandwidth of 500 MHz and allows the simultaneous use of a satellite by more than one ground station. Different ground stations are allocated different carrier frequencies within this 500 MHz band, either permanently or for particular periods of time − depending on the traffic originated by that station. This method of utilizing the capacity of a communication satellite is known as *frequency division multiple access* (fdma). Each of the allocated carrier frequencies has a sufficient wide bandwidth to allow a multi-channel telephony system to be transmitted and, in some cases, a television channel. The number of telephony channels thus provided varies from 24 in a 2.5 MHz bandwidth to 1872 in a bandwidth of 36 MHz. All the signals transmitted by a satellite are transmitted towards every ground station and each station selects the particular carrier frequencies allocated to it in the band 3.700−4.200 GHz.

13 Data Circuits

Many firms and organizations have a number of factories, warehouses, offices and other points where members of staff are employed. Usually, many of these locations are not sited at the headquarters, some indeed may be sited many hundreds or even thousands of kilometres distant. All of the remote locations need to send information to the headquarters at regular intervals varying from weekly to daily, or even more frequently. The information can, of course, be sent by post but there is an ever increasing demand nowadays by management for information to be made available more quickly. The answer to the problem can very often be given by telegraphic techniques, in particular the Telex system, but this is limited in its maximum speed of transmission, and some faster system is often required. Much of the data required by a head office is destined to be stored and/or processed by a digital computer, and computers are employed for a great number of varied purposes in both the engineering/scientific and the commercial fields. Computers are used to carry out complex calculations, to maintain and update records such as technical data, tax records and medical histories. Computers are also used in the preparation of weather reports and forecasts, the calculation and printing of wages and salaries, and of bills and invoices; they are also used for inventory control, and for the control of industrial processes in factories. The high-street banks make use of computers to maintain details of customers' accounts, of standing orders, direct debits, etc., and to operate cash machines, while airlines and package holiday firms are able to operate booking systems that provide an immediate confirmation of vacancies and bookings.

The many applications of the digital computer have led to most organizations investing in their own computer facilities. A main frame digital computer is an expensive purchase and it is not economically possible for an organization to install such computers at several locations. The facilities provided by a main frame computer may be needed at many different points in the organization's set-up. The computing facilities needed at different points may be provided by a minicomputer, or a microcomputer, but these computers sometimes may need to communicate with one another or with the main frame computer. Because of this there is a considerable demand for *data links*, to connect *data terminals* with the main *computer centre* and so extend the use of the main frame computer installed at the centre to a number of distant locations.

The Use of Leased Circuits and the PSTN

A data link that connects a *data terminal* to a remote digital computer may be leased from the telephone administration or it may be temporarily set up by dialling a connection via the *public switched telephone network* (pstn). The choice between leasing a private circuit and using the pstn must be made by the careful consideration of factors such as the cost, availability, speed of working, and transmission performance. Private circuits may transmit d.c. ± 6 V or ± 80 V signals or may use modems to convert the data signals into voice-frequency signals, or may transmit via a pcm system (p. 166). The maximum length of a ± 6 V data link is short but when ± 80 V signals are used almost any distance can be covered albeit at only a low bit rate (i.e. up to 150 bits/s). Voice-frequency data circuits can also be of any length and may be routed, wholly or partly, over loaded or unloaded audio-frequency cable, over multi-channel telephony systems, or over a radio link. (A multi-channel system may itself be routed, wholly or partly, over a radio link and may employ either fdm or tdm techniques.)

A point-to-point privately leased circuit provides exclusive use of the transmission path and so its parameters, such as its attenuation/frequency and group-delay/frequency characteristics, can be equalized and adjusted to give the optimum performance which enables high bit rates to be reliably transmitted. A pstn connection is established by dialling the telephone number of the distant data terminal and the route used for a particular call is a random choice from a large number of different possible routings, including various combinations of audio-frequency cable and multi-channel system. Because of this the overall attenuation and group-delay/frequency characteristics of a pstn connection are not predictable and so they cannot be adjusted to give the best transmission performance. Transmission at 2400 bits/s and 4800 bits/s is available over the pstn but the results cannot be guaranteed; some of the dialled connections may not be adequate and the error rate may be too high. When synchronous modems are employed bit rates of 9600 bits/s or more are commonly available over leased lines. One modem offers a bit rate of 14 400 bits/s, with reduced speed facilities at both 12 000 bits/s or 9600 bits/s.

The bandwidth provided by a leased circuit is 300—3000 Hz, the lines being adjusted to give a good transmission performance over this bandwidth. At higher frequencies group-delay/frequency distortion increases rapidly. The bandwidth made available by the pstn is more restricted, nominally 300—500 Hz and 900—2100 Hz, because of the need to avoid the frequencies used for in-band signalling systems on trunk routes. Signalling tones may also be present on some leased circuits.

A pstn connection is noisier than a leased circuit because switching equipment in telephone exchanges is a source of considerable noise and interference. The effect of noise and interference voltages is to reduce the maximum bit rate that can be employed for a given error rate.

The use of a privately-leased circuit for data transmission has the following advantages over the use of the pstn.

(*a*) Exclusive use of the circuit is obtained. Thus time is not wasted setting up calls and the performance of the circuit remains stable.

(*b*) The link can be adjusted to have the optimum performance and it is less affected by noise and interference. As a result higher speed and more reliable transmissions are possible.

(*c*) Full-duplex operation is available at higher bit rates and higher data 'throughput' is possible by eliminating turn-round time. (With half-duplex operation the time taken for the modem to switch from reception to transmission and vice versa — known as the *turn-around time* — is some tens of milliseconds.)

On the other hand, the cost of permanently leasing a line may be relatively high and it can only be economically justified if there is sufficient data traffic on the line or the particular terminal application necessitates a permanent connection, e.g. a cash dispensing terminal in a bank which checks a customer's account before releasing the cash. If the data communication requirements involve occasional contact with a large number of locations, and the majority of the connections are of fairly short duration, the use of the pstn is probably best. On the other hand, if long duration connections between a few branch offices and the computer centre is planned for, leased links will probably be chosen. In practice, most private data networks consist of a combination of both leased and pstn links, and very often the leased circuits are provided with the stand-by facility of using the pstn when necessary (i.e. if the leased circuit should fail).

Two-wire and Four-wire Presented Circuits

The terms two-wire and four-wire refer to the line circuit that is presented to the data terminal. Figure 13.1 shows the block diagram of a link that is operated two-wire over the whole of its length. Such an arrangement is only possible for a short line unless a low-bit speed ± 80 V circuit is satisfactory. Higher speeds of transmission require

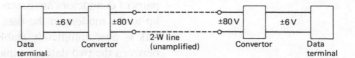

Fig. 13.1 Two-wire d.c. data circuit

the use of modems at each end of the line to convert the data into voice-frequency signals and Fig. 13.2 shows the circuit of an amplified data link. The line joining the two terminal repeater stations is operated on a four-wire basis for reasons discussed in Chapter 10, but, since the local lines connecting the data terminals to their nearest repeater stations are operated two-wire, the circuit as a whole is said to be two-wire presented. The four-wire section of the circuit is most likely to be routed over a channel in a multi-channel system. Increasingly, data is routed over pcm systems.

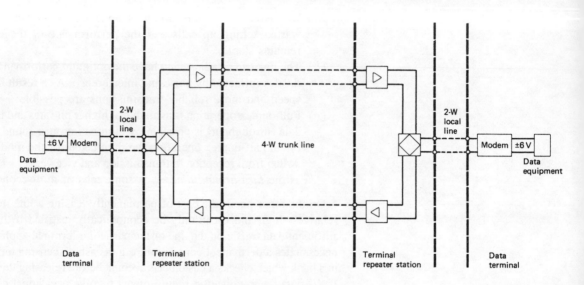

Fig. 13.2 Two-wire data circuit using modems and amplifiers

When fsk is employed the two-wire sections of the circuit are only able to transmit 600 bits/s or 1200 bits/s signals in one direction at a time (half-duplex) since the frequencies used are 1300 Hz and either 1700 Hz or 2100 Hz and these occupy most of the available frequency band. Full-duplex operation is only possible if the return direction of transmission is operated at a different frequency and at a lower bit rate. Alternatively, full-duplex operation is possible if both directions of transmission are operated at the lower speed of 300 bits/s. For the 300 bits/s duplex operation the CCITT recommended frequencies are given in Table 13.1. The two directions of transmission are operated in different frequency bands and so they can be sorted out at the receiver by means of suitable filters.

When a data link is operated on a four-wire presented basis, two pairs of conductors are extended from the local repeater station right up to the modem in the data terminal as shown by Fig. 13.3. The link is now completely stable and provides two separate channels between the two data terminals, each of which is able to carry high-

Table 13.1

| Channel | Frequency (Hz) | |
	Binary 1	Binary 0
1	980	1180
2	1650	1850

Fig. 13.3 Four-wire data circuit

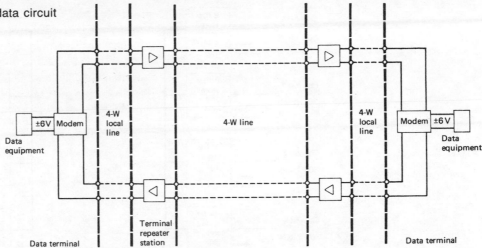

speed data simultaneously. Thus, the four-wire link provides full duplex high-speed operation.

Multiplexers, Front-end Processors and Concentrators

Because of the high costs of line plant it is desirable to be able to use a line to carry more than one data link. *Multiplexers* and *concentrators* are two equipments which, in their different ways, increase the utilization of a point-to-point circuit. Further savings in costs are obtained with a concentrator since fewer modems and interfacing equipments are then required. A front-end processor relieves the digital computer of many tasks and allows time-sharing of the computer.

Multiplexers

A **multiplexer** combines several individual data links together using either *frequency-division multiplex* (fdm) or *time-division multiplex* (tdm). With *frequency-division multiplex* each data channel is shifted, or frequency translated, to a different part of the available frequency spectrum. The particular frequency to which a channel is shifted is determined by the frequency of the carrier wave which is modulated by that channel. For example, a 300–3400 Hz bandwidth point-to-point circuit can accommodate 12 110 bits/s data channels. The bandwidth needed per channel is 55 Hz but, because of the need for interchannel frequency gaps to allow the filter to build up its attenuation, the carrier frequencies are spaced 240 Hz apart. This system is not often employed since tdm can multiplex more channels *and* operate at a higher bit rate.

With *time-division multiplex* the data channels each occupy the same

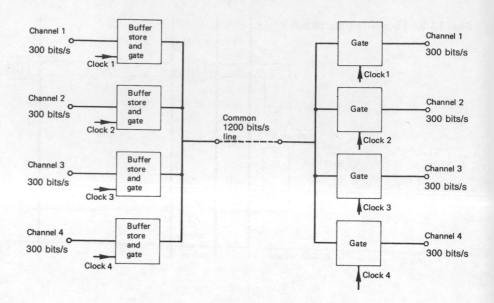

Fig. 13.4 Use of time-division multiplexers

frequency band but they are applied in time sequence to the line. Figure 13.4 illustrates the basic principle of a time-division multiplexer. Four data channels are shown operating at 300 bits/s. The duration of a bit is $\frac{1}{300}$ s or 3.33 ms and so a 7-bit character occupies a time slot of 23.31 ms. Suppose that the common line is to be operated at the higher speed of 1200 bits/s so that each bit sent to line will have a time duration of 0.83 ms. The data present on the channel 1, 2, 3 and 4 inputs to the tdm system are fed into the appropriate channel buffer stores and held there until the store is given access to the common line. The clock pulses 1, 2, 3 and 4 are applied to each gate in turn to sample the stored information held in the associated buffer store and apply it to the common line for a time period equal to the duration of a character, i.e. 23.31 ms. When, for example, clock pulse 1 enables store A for 23.31 ms, the stored data is transmitted to line at the 1200 bits/s bit rate. Clock pulse 1 then ends and buffer store A is taken off line. Clock pulse 2 now enables store B to allow one character of the data stored to be sent to line. After 23.31 ms buffer store B is inhibited and buffer store C enabled and so on until all four channels have been connected, in turn, to the common line. Clock pulse 1 now enables buffer store A again and another character of the stored data can be transmitted to the line and so on. Character-interleaving is used in conjunction with start—stop synchronization although the start—stop bits are often *stripped* whilst the data is transmitted over the mux link.

Clearly, synchronization between the transmitting and receiving equipments is essential in order that the clock pulses at the receiver occur at the correct intervals in time. Otherwise, the pulses proper to one channel may well be routed to another channel. Bit interleaving is also possible but is only employed with higher-speed synchronous systems.

Usually, the output of a tdm equipment will be applied to a modem to convert the digital signals into voice-frequency signals for transmission over the telephone network (see Fig. 13.5). With a tdm system the bit rate on the common highway is the *sum* of the individual channel bit rates.

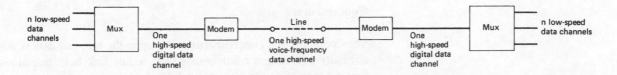

Fig. 13.5 Multiplexed voice-frequency data circuit

EXAMPLE 13.1

A 2400 bits/s data link is to carry one 1200 bits/s and a number of 300 bits/s data channels using time-division multiplex. Determine the number of 300 bits/s channels that can be transmitted.

Solution
Line bit rate = sum of channel bit rates = 2400 bits/s.
Hence, number of 300 bits/s channels = 1200/300 = 4.

Front-end Processors

A large digital computer generally interfaces with a data communication system through a **front-end processor** (fep). The fep relieves the computer of some tasks it would otherwise have to perform and enables its computing power to be devoted to the processing and storage of data. The fep performs the following tasks.

(*a*) The fep acts as a *multiplexer* to allow several data channels to have access to the computer on a time-sharing basis. One or more high-speed channels connect the fep to the computer, while the other side of the fep is connected to a greater number of data links.

(*b*) The fep acts as a communications controller. It controls all the telecommunications facilities of the computer, monitoring all the associated modems to determine when any of them has data ready for processing. The fep decides when a modem shall have access to the computer itself and so ensures that the computer is not overloaded by a large number of messages arriving at the same time. The fep sets up all the necessary connections via the pstn and meters all incoming calls and then produces bills for the use made of the computing facilities.

(*c*) The fep *polls* or asks each data terminal in turn whether it has any data to transmit to the computer. The interrogated terminal can reply by sending its data or it can signal that it has none to send. The fep can then check whether it has any data to send to that data terminal and if so transmits it. The fep will then poll the next terminal in the laid-down sequence. The polling can be carried out at such a speed that the fep acts as a multiplexer to provide time-sharing of the computer.

Concentrators

The operation of a **concentrator** relies upon the fact that data is not normally transmitted continuously over a data link but, instead, is sent in *bursts* of varying lengths. A line concentrator is able to connect any of x inputs to any of y outputs, where $x > y$. The idea is illustrated by Fig. 13.6 in which any input line a, b, c, d or e can be connected to any output line A, B or C. The concentrator polls each of its inputs in turn to discover whether it has data waiting to be transmitted. If it has, that input is switched through to an unoccupied output line and the distant modem is signalled that data transmission is about to commence. When incoming data is received via one of the lines A, B or C, the concentrator determines the destination data terminal and addresses the data.

Fig. 13.6 Use of a line concentrator

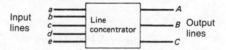

Data concentrators include storage facilities which allow the input data to be stored and then read out from the store and transmitted down the line in blocks when the concentrator is polled by the fep at the computer centre. The storage facility allows data to be sent in smooth blocks or trains of data (which may well be the combination of the data originating from more than one terminal) and allows the use of a main line whose bit rate is less than the sum of the input bit rates. This does mean, of course, that if all the input lines are sending data at the same time, some of the data will have to be stored and there will be some delay in the concentrator before all of the data is passed on to the main line. Figure 13.7 shows how a data concentrator is used; it should be noted that the concentrator works with digital signals only and any analogue signals must be changed into ± 6 V signals before entering the circuit.

Sometimes, a line concentrator is used before a multiplexer to achieve the maximum utilization of the transmission path and Fig. 13.8

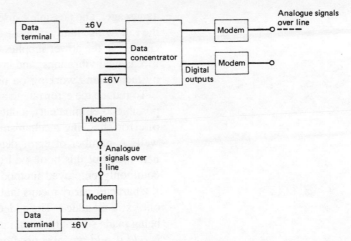

Fig. 13.7 Use of a data concentrator

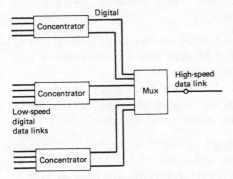

Fig. 13.8 Use of concentrators and multiplexers

gives an example of this technique. For each of the line concentrators the bit rate at its output is less than the total bit rate at its input but for the multiplexer the output bit rate is higher than the input bit rate on each input.

A multiplexer can also be used as a simple kind of front-end processor to permit time-sharing of a digital computer and Fig. 13.9 shows a possible data network. The fep polls each of the concentrators sequentially and during the short time a concentrator is connected to the computer it can pass data into, or receive data from, the computer. Clearly, the concentrator must possess some storage facility so that it can store the data received when the computer is dealing with another concentrator, or a direct link. As shown in the figure some concentration may be provided at the computer centre but other concentrators may be located considerable distances away.

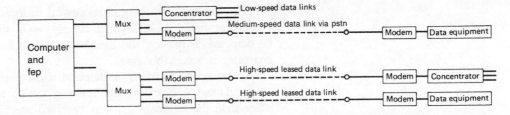

Fig. 13.9 A data network including both multiplexers and data concentrators

Error Control

A data link is subjected to a number of sources of noise which can produce errors in the received data. More serious are the errors produced by brief breaks in the transmission path; the loss of a single bit may well cause an error which could alter the whole meaning of a message. The possible causes of interruptions are many and include the following: faulty or loose test connections, poorly soldered joints, changes in the power supplies from the main plant to the standby plant, mechanical vibrations, and last, but by no means least, breaks caused by technicians working on nearby circuits or equipments.

To reduce the error rate to an acceptable level (it is not economically possible to eliminate it), a data system will usually include some kind of error-checking mechanism and in some cases error correction as well. A number of error detection methods are in use in different networks but this book will only outline the principles of the most commonly employed method, known as the *parity system*.

Character parity means that each character signalled to the line has one extra bit added. The added bit may be either a 1 or a 0; the choice being made so that the *total* number of 1 bits transmitted per character is *odd* if *odd parity* is used or *even* if *even parity* is employed. The parity bit is transmitted at the end of each character. Two examples of this system are given below.

(*a*) Character 0 0 1 1 1 0 1; there are four 1 bits in this character and so the added parity bit must be a 1 if odd parity is used and a 0 if even parity is employed.

(*b*) Character 1 0 1 0 1 1 1; thre are five 1 bits in this case and so odd parity requires the addition of a 0 parity bit, while if even parity is used the added parity bit must be a 1.

At the receiver, the incoming character is checked to determine how many 1 bits it contains, and if the total is odd (or even) the received character is taken as being correct. Should the total number of 1 bits received be an even number when odd parity is used, or an odd number when even parity is used, an error in the received message is detected which must, in some way, be indicated at the receiving terminal or signalled back to the transmitter so that the data can be sent again.

Parity bit error checking is successful in locating any single-bit errors that should occur although clearly two bits in error would not be detected. However, the likelihood of two bits in error in one character is acceptably small.

Two-bit errors can also be detected if the parity bit principle is extended to whole blocks of data. It is also possible to use parity checks other than the number of 1 bits received per character but such procedures are beyond the scope of this book. In practice, block parity systems are normally used and character parity is fairly rare.

With an *error correction* system it is usual to acknowledge each character or each block as it is received. Character-by-character acknowledgement is very time consuming when the error rate is low but, on the other hand, block error correction means repeating a complete block of data whenever an error does occur. For either error correction method, the next character of the next block is not sent by the transmitter until an Ack. signal is received from the distant receiver.

Typical Data Networks

A data network will contain one or more digital computers and a number of data terminals, each of which will have some kind of access to a computer. The layout of a possible system is shown by Fig. 13.10. The network shown is a composite of the various techniques outlined earlier in this chapter.

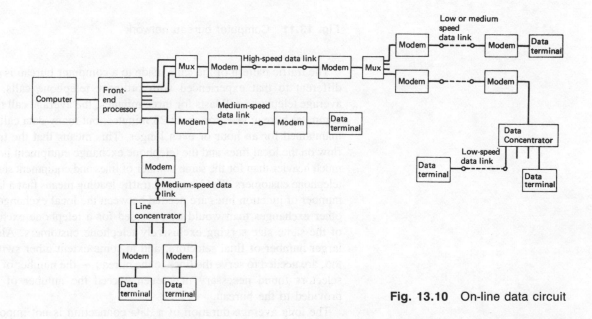

Fig. 13.10 On-line data circuit

Many smaller businesses may have a need for the facilities provided by a main frame computer but be unable to economically justify the purchase of one. To meet this demand *computer bureaux* rent computer time to their customers. Figure 13.11 shows how a bureau network might be set up. A number of data terminals are able to operate simultaneously (apparently) with the remote computer using the pstn or, in some cases, a privately-leased circuit. The network makes computing power available, on request, to a large number of customers located anywhere in the network.

When a link is to be established over the pstn the caller presses the TELE button on his telephone receiver and then dials the number of the computer bureau. Once the call has been answered, the DATA button on the telephone is pressed and control of the connection passes to the modem. The data number of the caller is then transmitted to *log-in* to the computer. Once the computer has signalled that the logging-in is satisfactory, the transmission of data can begin.

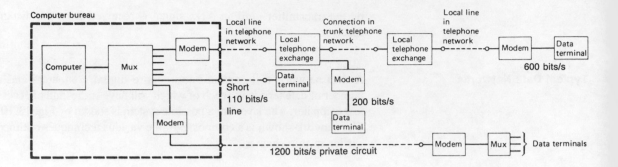

Fig. 13.11   Computer bureau network

The traffic pattern of the calls made to a computer bureau is quite different to that experienced with ordinary telephone calls. The average telephone call lasts for three minutes, the average call to the computer bureau lasts for about 25 minutes, but many data calls are maintained for an hour or even longer. This means that the traffic flow on the local lines and the telephone exchange equipment is very much heavier than for the same number of lines and equipment serving telephone customers. The increased traffic loading means that a larger number of junction lines are needed between the local exchange and other exchanges than would be provided for a telephone exchange of the same size serving exclusively telephone customers. Also, a larger number of final selectors, and to some extent other switches too, are needed to serve the lines to the bureau — the number of final selectors found necessary may even exceed the number of lines provided to the bureau.

The long average duration of a data connection is not important to a user when only a local telephone call is needed to gain access to the computer bureau but it could prove to be expensive when long-distance trunk calls are involved. To overcome this problem a multiplex (frequency-division or time-division) system can be provided on a leased private circuit from the computer bureau to a convenient remote (from the bureau) point. The customer at the remote point can then make a local call to gain access to his end of the multi-channel system and thence to the bureau.

One of the more common kinds of data network is the kind where data originating from a large number of variously located branch offices is to be processed by a central computer, stored, and when needed is transmitted back to the originating or some other branch office. This is, of course, the type of network used by the high-street banks. Each branch office has access to a central computer which keeps details of customers' accounts, standing orders, direct debits, and so on, and prints out bank statements, and pays standing orders and direct debits by transferring data from one location to another. When a payment is made from an account with one bank to an account held with another bank, a computer-to-computer transfer of data will be needed.

Figure 13.12 shows the layout of a possible centralized accounting data system. The branching points allow a number of data terminals to share a common line without any change in the bit rate, the terminals are polled by the fep and are signalled when they can transmit their data. Should a branch be unable to establish contact with the computer over its direct route for some reason, a link can be set up instead over the pstn.

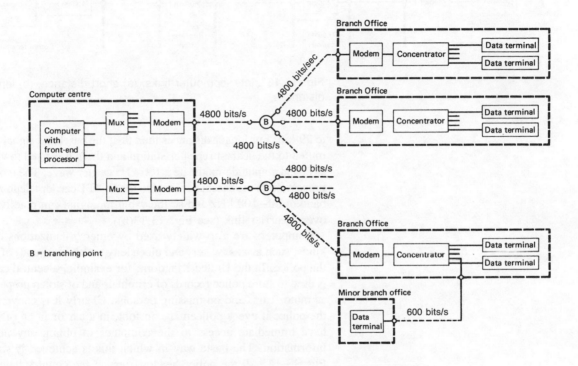

Fig. 13.12   Centralized accounting data network

When two computers are to communicate directly with one another they can be directly linked together without interface equipment only when the distance between them is very short. For longer distances the arrangements shown in Fig. 13.13(a) or Fig. 13.13(b) can be employed. In Fig. 13.13(a) the modem encodes the data outputed by the computer into a random pattern. This is because repetitive data patterns will have large-amplitude components in their frequency spectra and these may cause interference with other pairs in the cable. The random data pattern generated by the modem is band-limited to 1.92−36 kHz (known as the *baseband*). The maximum distance that can be linked in this way is limited, by the line attenuation at 36 kHz,

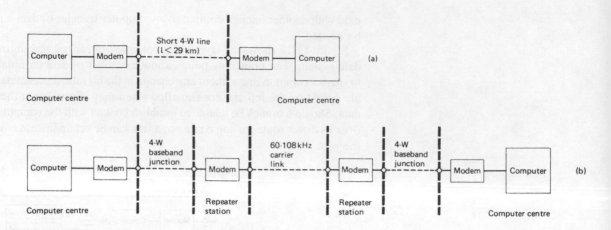

Fig. 13.13 Inter-computer links: (a) short distance, (b) longer distances

to 29 km. For longer distances than this, the baseband signal is transmitted to the nearest repeater station and there it is used to vestigial-sideband amplitude-modulate a 100 kHz carrier wave. The modulation process shifts the baseband signal to the CCITT carrier telephony baseband of 60–108 kHz and so the vsb a.m. signal can readily be sent over a carrier link (see Fig. 13.13(b)).

Computers are also widely used by other organizations of many kinds, such as gas, water, and electricity authorities, and, of course, the police. In the United Kingdom, for example, a central computer is used to store police records of criminals and of stolen property such as motor cars, and on missing persons. Clearly it is convenient for the police if every policeman, on foot, in a car or in an office, can have immediate access to the computer to obtain any necessary information. The basic way in which this is achieved is shown by Fig. 13.14. All the police headquarters in the country have access to the centralized police computer and all policeman have access to a data terminal. This access may be via a radio link or a telephone call as shown in the figure or, when it is convenient, by personal contact with the data operators.

Data Circuits over PCM Systems

The CCITT pcm system provides thirty 64 kb/s channels with the bearer circuit operating at 2.048 Mb/s. These bit rates are much larger than the speeds the analogue network, especially the pstn, can handle. This means that the standard pcm system has the capability to accommodate a number of high-speed data circuits. Each terminal of a speech or analogue pcm system includes circuitry for analogue-to-digital conversion (adc) and for digital-to-analogue conversion (dac). When, therefore, one or more of the thirty channels is to carry a data circuit the data signal is converted to voice-frequency before it is applied

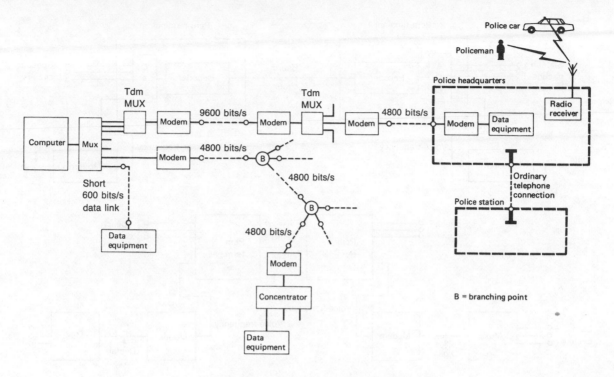

Fig. 13.14   Police computer network

to the pcm channel input. This step is carried out so that the terminals need not be modified by having some of their adc and dac equipments removed. Thus, so far as the pcm system is concerned, it is only carrying a number of voice-frequency analogue signals in digital form. The arrangement employed is shown by Fig. 13.15(a). The 64 kb/s channel provided can be multiplexed to provide an increased number of lower-speed channels, see Fig. 13.15(b).

If the entire 2.048 Mb/s capacity of a pcm system is to be used for the transmission of digital signals the pcm terminal equipment that is employed is *not* fitted with either channel adc or with channel dac. The input digital signals are directly applied to the encoder and converted into the pcm format. At the receiving end of the system the pcm signals are directed to their appropriate channels, decoded, and then outputted in digital form. Figure 13.16 shows the basic principle of the *digital* pcm system; each of the 64 kb/s channels may be subdivided into a number of lower-speed channels by digital multiplexing equipment. This equipment can be provided at the site of the pcm terminal or at the premises of the user. British Telecom offers digital services which are based upon the method given in Fig. 13.16.

(*a*) *Kilostream* is a digital point-to-point service which is available at bit rates of 2400 b/s, 4800 b/s, 9600 b/s, 48 kb/s and

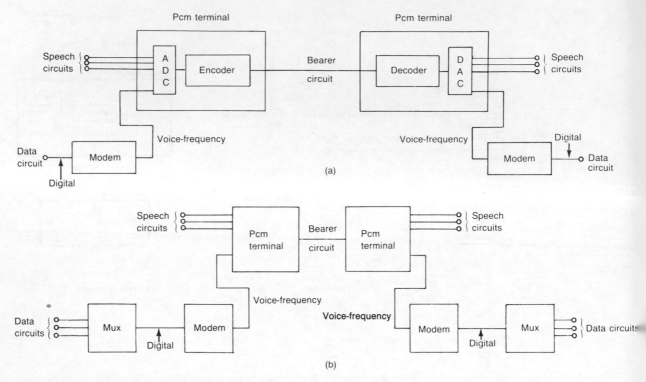

Fig. 13.15 Data signals routed over one channel in a pcm system

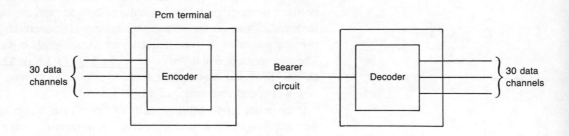

Fig. 13.16 Digital pcm system

64 kb/s. The service is provided using a 64 kb/s channel and the appropriate multiplexing equipment (except 64 kb/s). At 2.4, 4.8 and 9.6 kb/s kilostream offers a cheaper alternative to analogue transmission using modems.

(b) *Megastream* offers a point-to-point digital service operating at either 2.048 Mb/s or at 8 Mb/s. The service can be used without multiplexing to provide a wide-band high-speed system or it can be multiplexed to give a number of lower-speed channels. The channels need not all carry data; if required some of the channels can carry digitized speech signals instead.

(c) *Switchstream* offers full-duplex data communication at speeds of up to 48 kb/s. It employs a technique that is known as packet switching.

(d) *Satstream* provides leased high-speed links via communication satellites to countries in Western Europe.

The use of a digital circuit to carry data signals offers a number of advantages.

(a) Impulse noise is eliminated.

(b) White noise is not cumulative.

(c) Because of (a) and (b) data can be transmitted at higher bit rates.

(d) An expensive modem is not needed at each end of the data link.

14 Optical Fibre Systems

Visible and infrared light extends over a range of wavelengths of about $0.4~\mu$m to about 1 mm but the use of fibre optics is, at present, restricted to the approximate waveband of $0.8~\mu$m to $1.6~\mu$m. The energy carried by light waves in this waveband can be transmitted by glass fibres which act as *dielectric waveguides*. Essentially, an **optical fibre** consists of a cylindrical glass *core* that is surrounded by a glass *cladding*. The use of an optical fibre to transmit light energy offers a number of advantages over the more conventional transmission techniques using copper or aluminium conductors. These advantages are as follows.

(*a*) Light-weight, small-dimensioned cables.
(*b*) Very wide bandwidth.
(*c*) Freedom from electromagnetic interference.
(*d*) Low attenuation.
(*e*) High reliability and long life.
(*f*) Cheap raw materials.
(*g*) Negligible crosstalk between fibres in the same cable.

Fibre optic technology is now well established as an alternative to copper cable for use in long-distance, wideband communication systems. Optical fibre is particularly suited to the transmission of digital signals, and the most important telecommunications applications are in connection with pcm multi-channel telephone systems. Many other applications of fibre optics exist in other fields; for example, automobile electronics and industrial control systems.

Propagation in an Optical Fibre

When a light wave travelling in one medium passes into another medium, its direction of travel will probably be changed. The light wave is said to be *refracted*. The ratio

$$\frac{\text{Sine of angle of incidence}}{\text{Sine of angle of refraction}} = \frac{\text{sine } \varphi_i}{\text{sine } \varphi_r} \quad \text{(see Fig. 14.1)}$$

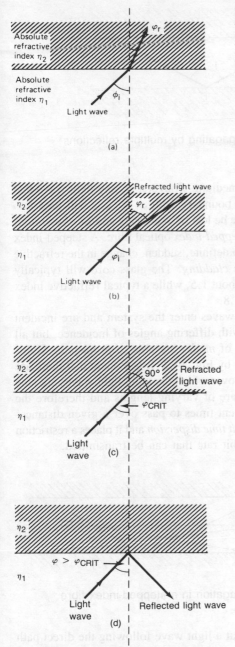

Fig. 14.1 Reflection and refraction at the boundary of two media with differing refractive indices

is a constant for a given pair of media. This constant is known as the *refractive index* η for the two media. If one of the two media is air, the *absolute refractive index* of the other medium is obtained. If the absolute refractive indices of the two media are η_1 and η_2 respectively, then

$$\sin \varphi_i / \sin \varphi_r = \eta_2 / \eta_1$$

When the light passes from a medium of higher refractive index into a medium of lower refractive index, i.e. when $\eta_2 > \eta_1$, the wave will be bent towards the normal (Fig. 14.1(a)). Conversely, if $\eta_1 > \eta_2$, the light wave will be refracted away from the normal (see Fig. 14.1(b)). Some of the incident light energy will be reflected at the boundary of the two media. As the angle of incidence is increased, the angle of refraction will also be increased and the point will be reached at which $\varphi_r = 90°$. The refracted wave will then travel parallel to the boundary of the two media (Fig. 14.1(c)). If the angle of incidence is further increased the light wave will be *totally reflected*. The angle of incidence at which total reflection first occurs is known as the *critical angle* φ_{crit} (see Fig. 14.1(d)). The value of φ_{crit} depends on the absolute refractive indices of the two media according to the equation

$$\varphi_{crit} = \sin^{-1} \frac{\eta_1}{\eta_2} \tag{14.1}$$

The angle of reflection is always equal to the angle of incidence.

EXAMPLE 14.1

Light is passed through glass of refractive index 1.5. Calculate the angle of incidence at which total internal reflection will take place.

Solution
For total internal reflection $\varphi_r = 90°$ and so $\sin \varphi_r = 1$.
   Therefore,

$$1.5 = \frac{1}{\sin \varphi_i}$$

$$\sin \varphi_i = \frac{1}{1.5} = 0.667$$

$$\varphi_i = \sin^{-1} 0.667 = 41.8° \quad (Ans.)$$

Suppose that now another medium, known as the *cladding*, also of refractive index η_2, is placed on the other side of the lower medium, as shown by Fig. 14.2. Provided the angle of incidence φ_1 of the input light wave is larger than the critical value φ_{crit}, the light wave will be able to propagate along the inner medium, or *core*, by means of a series of total reflections. Any other light waves that are also incident on the upper boundary at an angle $\varphi_1 > \varphi_{crit}$ will also

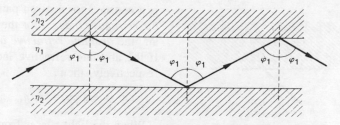

Fig. 14.2 Light wave propagating by multiple reflections

propagate along the inner medium. Conversely, any light wave that is incident upon the upper boundary with $\varphi_1 < \varphi_{crit}$ will pass into the upper medium and there be lost by *scattering* and/or *absorption*.

This is an example of a *stepped-index* optical fibre. A stepped-index optical fibre is one that has a definite, sudden, change in the refractive indices of the *core* and the *cladding*. The glass core will typically have a refractive index of about 1.5, while a typical refractive index for the cladding is about 1.8.

When a number of light waves enter the system and are incident upon the upper boundary with differing angles of incidence, but all greater than φ_{crit} a number of *modes* are able to propagate. Multimode propagation is shown by Fig. 14.3. It is evident that there are a number of different paths over which the input light waves are able to propagate. These paths are of varying lengths and therefore the light waves will take different times to pass over a given distance. This effect is known as *transit time dispersion* and it places a restriction on the maximum possible bit rate that can be transmitted.

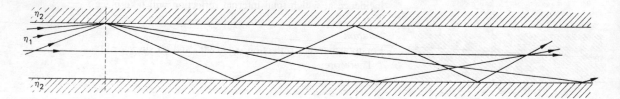

Fig. 14.3 Multimode propagation in a stepped-index fibre

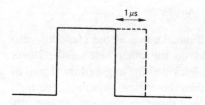

Fig. 14.4 Effect of transit time dispersion on a rectangular pulse

Suppose, for example, that a light wave following the direct path takes 5 μs to travel a certain distance, and that a wave travelling over the longest path takes 6 μs to reach the same point. The effect on a rectangular pulse transmitted over the system is shown by Fig. 14.4. The trailing edge of the pulse is delayed in time by 1 μs. This effect will, of course, limit the maximum bit rate that is possible, since the leading edge of a following pulse must not arrive before the extended edge of the pulse shown has ended.

Dispersion in an optical fibre can be minimized in one of two ways.

(a) A *graded-index fibre* can be used. The refractive index of the core is highest at the centre and then decreases gradually towards the edges. This ensures that the difference between the refractive indices of the core and the outer medium or cladding is small, and the change can take place smoothly instead of abruptly as before. Light waves nearing the boundary of the two media will then be gradually refracted, rather than reflected, from the boundary as shown by Fig. 14.5. A wave entering the core at a large angle from the horizontal will penetrate a long way from the horizontal before it is refracted sufficiently to change its direction of travel. A wave entering the core at a shallower angle does not penetrate as far from the horizontal before it is refracted sufficiently to change its direction of travel.

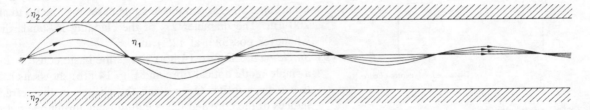

Fig. 14.5 Multimode propagation in a graded-index fibre

Once a wave has been refracted back to the horizontal, it will enter the next section of the core at a shallower angle than before and it will not, therefore, travel as far from the horizontal before reversing its direction of travel. The effect of this is to reduce the differences between the lengths of the paths followed by the various light waves. Also, the waves that travel farthest from the axis are travelling in a lower refractive index medium and so travel at a higher velocity than waves travelling near the axis. These effects ensure that all the light waves travelling through the core have almost the same transit time.

(b) A *single-mode* or *monomode fibre* can be used. If the diameter of the inner medium or core is reduced to be of the same order of magnitude as the wavelength of the incident light wave, then only one mode will be able to propagate (see Fig. 14.6).

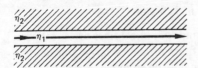

Fig. 14.6 Monomode propagation in a stepped-index fibre

No matter which of the three possible modes of propagation is used, the dimensions of the outer medium or cladding must be at least several wavelengths. Otherwise some light energy will be able to escape from the system, and extra losses will be caused by any light scattering and/or absorbing objects in the vicinity. The available bandwidth is limited by (a) multimode dispersion and (b) material dispersion.

Fibre Optic Cable

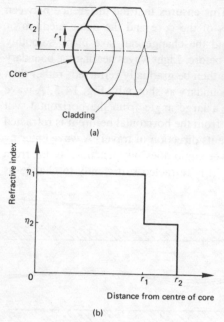

(a)

(b)

Fig. 14.7 (a) Stepped-index multimode optical fibre, (b) Refractive index profile

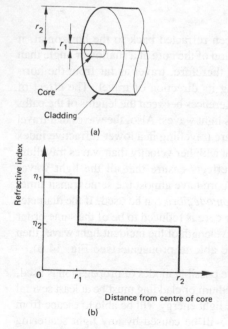

(a)

(b)

Fig. 14.8 (a) Single-mode optical fibre, (b) refractive index profile

An optical fibre cable consists of a glass core that is completely surrounded by a glass cladding. The *core* performs the function of transmitting the light wave(s), while the purpose of the *cladding* is to minimize surface losses and to guide the light waves. The glass used for both the core and the cladding must be of very high purity since any impurities that are present will cause some scattering of light to occur. Two types of glass are commonly employed: silical-based glass (silica with some added oxide) and multi-component glass (e.g. sodium borosilicate).

There are three basic versions of an optical fibre cable.

(a) *Stepped-index Multimode*. The basic construction of a stepped-index multimode optical fibre is shown by Fig. 14.7(a) and its refractive index profile is shown by Fig. 14.7(b). It is clear that an abrupt change in the refractive index of the fibre occurs at the core/cladding boundary. The core diameter, $2r_1$, is often some $50-60$ μm but in some cases may be up to about 200 μm. The diameter $2r_2$ of the cladding is standardized, whenever possible, at 125 μm.

(b) *Single-mode*. Figure 14.8(a) shows the basic construction of a single-mode optical fibre and Fig. 14.8(b) shows its refractive index profile. Once again the change in the refractive indices of the core and the cladding is an abrupt one but now the dimensions of the core are much smaller. The diameter of the core should be of the same order of magnitude as the wavelength of the light to be propagated and it is therefore in the range $1-10$ μm. The cladding diameter is the standardized figure of 125 μm.

(c) *Graded-index Multimode*. The basic construction of a graded-index multimode optical fibre is the same as that of the stepped-index multimode fibre shown in Fig. 14.7(a). The refractive index profile is not the same however and this is given by Fig. 14.9. The core has its greatest value of refractive index at its centre and this decreases parabolically towards the boundary with the cladding. The core diameter is in the range $50-60$ μm and the cladding diameter is 125 μm.

The relative merits of the three kinds of optical fibre are as follows. The stepped-index multimode fibre produces large transit time dispersion and in consequence it can only be employed for *bandwidth-distance products* of up to about 50 MHz km. The use of this kind of optical fibre is therefore restricted to applications such as low-speed data signals and various industrial control systems.

Single-mode optical fibre can provide very large bandwidth-distance products, of up to about 100 GHz km. Early difficulties in manufacture and jointing, and in the manufacture of coupling devices led to the development and use of graded-index optical fibre. Now these difficulties have been overcome and single-mode fibre is increasingly employed.

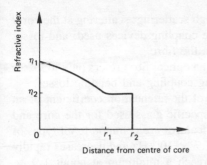

Fig. 14.9 Refractive index profile of graded-index multimode optical fibre

Graded-index multimode optical fibre has the ability to give bandwidth-distance products of 1 GHz km or more and, because of its larger dimensions, it is relatively easy to manufacture and install.

Glass is a brittle material and an optical fibre cable must be protected against breakage both during its installation and over its anticipated lifetime. The necessary strength is provided by a steel wire situated in the middle of the cable and protection is given by a plastic sheath. The constructional detail of one type of cable is shown by Fig. 14.10. The steel wire at the centre of the cable has a bedding layer around it and then the six optical fibres are laid helically around the layer. The gaps between the fibres are filled up with a filling compound and then an aluminium-tape water-barrier is wrapped around to form the cable. Lastly, a polyethylene sheath is placed around the aluminium tape to complete the cable.

In some cases copper conductors are also provided within the cable; they may be used for power-carrying purposes and for speaker circuits. An example of such a cable is shown in Fig. 14.11.

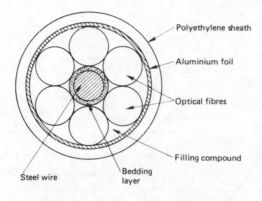

Fig. 14.10 Optical fibre cable

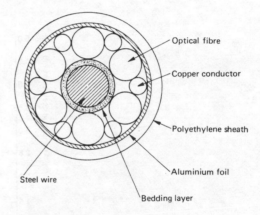

Fig. 14.11 Optical fibre cable with added copper conductors

Attenuation of an Optical Fibre

The light power output of the light source has to feed into the optical fibre with the maximum possible efficiency. To this end, *couplers* have been designed to maximize the *launching efficiency*. The launching efficiency is the ratio

$$\eta = \frac{\text{Power accepted by fibre}}{\text{Power emitted by light source}} \times 100\% \tag{14.2}$$

The light energy fed into an optical fibre is attenuated as it travels towards the far end. The losses in an optical fibre are contributed to by a number of sources. The sources of loss are: absorption; scattering in the core because of inhomogeneities in the refractive index — this is known as Rayleigh scattering; scattering at the core/cladding boundary; losses at the coupling devices used; and losses due to radiation at each bend in the fibre.

The attenuation coefficient of an optical fibre refers only to losses in the fibre itself, i.e. neglecting coupling and bending losses.

The variation with frequency of the attenuation coefficient of an optical fibre depends upon the specific glass used for the core and the cladding. However, certain features are common to all types of fibre. The loss is high at about 1.7×10^{14} Hz and decreases rapidly with increase in frequency to reach a minimum at about 1.9×10^{14} Hz. Further increase in the frequency results in a gradual increase in the attenuation coefficient up to about 5×10^{14} Hz. However, some sharp peaks in the attenuation coefficient exist at approximately 2.16×10^{14} Hz and 2.42×10^{14} Hz.

Because of the very high frequencies involved it is customary to work in terms of wavelength rather than frequency. Figure 14.12 shows typical attenuation/wavelength curves for both single-mode and graded-index multimode optical fibres. The peaks at 1.24 μm and 1.39 μm occur because of excess absorption loss at these wavelengths.

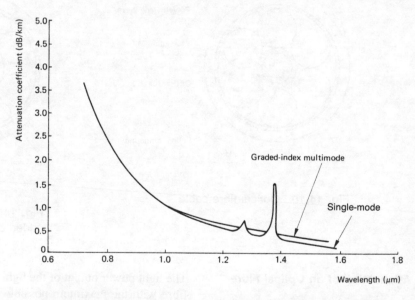

Fig. 14.12 Attenuation/frequency characteristics of single-mode and graded-index multimode optical fibres

Optical Sources

The light source used in an optical fibre system must, of course, be able to deliver light energy at the appropriate wavelength. It must also be capable of being modulated at the very high bit rates used, with a low driving power and a relatively high output power. There are two kinds of light source which are able to satisfy these requirements; these are the *light emitting diode* and the *laser diode*.

The light emitting diode (led) is generally used for short distance, narrow bandwidth systems, while laser diode finds application in long-haul wide-bandwidth systems. The laser diode can develop about 100 times more output power than the led can provide but it is both more expensive and less reliable at the present time.

The Light Emitting Diode

A semiconductor diode, fabricated using materials such as gallium arsenide (GaAs), gallium aluminium arsenide (GaAlAs), gallium indium arsenide phosphide (GaInAsP) or gallium indium arsenide (GaInAs), will emit light energy whenever it is forward biased and conducting a current. The emitted light is in the infrared part of the electromagnetic spectrum. The specific wavelength at which a **light emitting diode** radiates energy can be selected by the suitable choice of the semiconductor material (see Table 14.1).

Table 14.1

Material	GaAlAs	GaAs	GaInAs	GaInAsP
Wavelength of emitted light (μm)	0.8–0.9	0.9–1.0	1.0–1.1	1.25–1.35

There are two main types of led available, one of which radiates the light energy from the surface of the device whilst the other radiates from the edge of the semiconductor structure. The led may be used as the light source for telecommunication systems and produces non-coherent light output with a power of some 0.05 mW to 1 mW. The light is not of one wavelength but exhibits a *linewidth* or *spectral spread* that might be as much as 40 nm. Because of its non-coherent light output, the led will emit light waves into an optical fibre at a variety of angles and it is therefore suitable for use with graded-index multimode fibres. There is a linear relationship between drive current and optical power output.

The Laser Diode

The construction of a **laser diode** differs from that of a led mainly in that it incorporates a cavity which is resonant at the required frequency of the emitted light. When a voltage is applied across a laser diode, it radiates *coherent* light energy. The spread in wavelengths is small, perhaps only 1–2 nm. Commonly, the semiconductor material employed is gallium aluminium arsenide and this gives a peak emission at a wavelength of 0.82 μm with more than 100 MHz bandwidth. The power output is in the range of 1 mW to 10 W. Some commercially available laser diodes incorporate optical negative feedback to ensure a stable output over a wide range of ambient temperatures.

The laser diode offers the following advantages over the led: (*a*) small dimensions, (*b*) high efficiency, (*c*) it is easy to modulate, (*d*) directional emission, (*e*) output power ranges from a few milliwatts to about 10 W, and (*f*) a very short risetime (typically about 1 ns). On the other hand, the laser diode is more expensive, less reliable, and its performance is temperature sensitive. The laser diode finds its main applications in single-mode fibre systems operating mainly at a wavelength of 1.3 μm.

Optical Detectors

The function of an **optical detector**, or photo-detector, is to convert input light energy into the corresponding electrical signal. The detector should have its maximum detection efficiency at the operating wavelength of the system and should operate linearly at the modulated bit rate. Also, of course, it should be of physically small dimensions, and be cheap and reliable. All of these requirements can be satisfied by the semiconductor *photodiode*.

When a reverse-biased p-n junction is illuminated by light energy, a current will flow across the junction, whose magnitude is very nearly directly proportional to the illumination intensity. The current flowing is very nearly independent of the actual bias voltage so long as it keeps the junction reverse biased. If the diode current is passed through a load resistance, a detected output voltage can be obtained.

Two kinds of photodiode are employed: the silicon *pin photodiode* and the silicon *avalanche photodiode*. The pin photodiode is so named because it has a region of intrinsic semiconductor sandwiched in between n-type and p-type regions. This device is suitable for the shorter wavelengths but, for wavelengths longer than about 1.1 μm, its sensitivity is no longer good enough and then either a GaInAs or a GaInAsP device would be employed instead. The silicon avalanche photodiode is essentially a pin device also but it is fabricated in such a way as to accentuate the *avalanche effect*. The avalanche photodiode can provide a large current gain as well as detection but it has a tendency to be rather noisy.

Optical Fibre Telecommunication Systems

The basic block diagram of an optical fibre telecommunication system is shown by Fig. 14.13. The electrical signal to be transmitted over the system is fed into the optical transmitter. Here it is applied to the modulator and modulates the light source. The modulated light

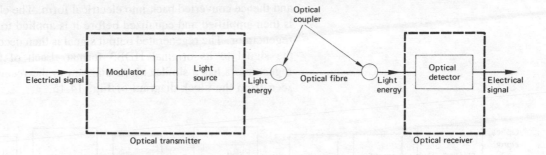

Fig. 14.13   Basic block diagram of an optical fibre system

energy is coupled to the optical fibre and is then transmitted to the far end using one of the modes of propagation described earlier. At the optical receiver, incoming light energy is detected in order to convert the signal back to its original electrical format.

The characteristics of the light source make it best suited to some form of digital modulation in which the source will be switched ON and OFF. The potential bandwidth offered by an optical fibre is very wide and usually the input electrical signal consists of the output from a pcm-tdm system. For example, 120 channels can be transmitted at a bit rate of 8 Mbits/s. 1920 channels can be transmitted at 140 Mbits/s or the equivalent of 8000 channels at 565 Mbits/s. Older systems operate at a wavelength in the neighbourhood of 0.84 μm but now the British Telecom standard is 1.3 μm over single-mode cable.

The HDB3 coded output signal of the pcm system is not suitable for transmission over an optical fibre and must be converted into a more suitable format before transmission. Figure 14.14 shows the block diagram of a suitable arrangement. The HDB3 signal is scrambled to produce a purely random pulse train and thus avoid any possibility of repetitive patterns occurring (these would tend to cause interference with adjacent optical fibres).

The encoded signal is then applied to the modulator and this modulates the laser diode. The light signal produced by the diode is then coupled into the optical fibre cable. At one, or more, points along the route the optical digital signal is regenerated to restore the shape of the signals.

At the receiver the incoming light signal is detected by the diode

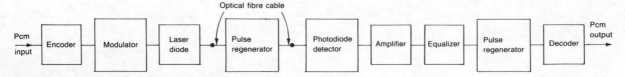

Fig. 14.14   System with optical pulse regenerator

and thence converted back into electrical form. The electrical signal is then amplified and equalized before it is applied to another pulse regenerator. The regenerated output signal is then decoded to restore the signal to its original HDB3 format. Each of the line pulse regenerators has a similar block diagram to the receiver, as can be seen from the block diagram of Fig. 14.15.

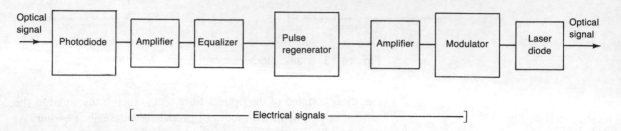

Fig. 14.15 Optical pulse regenerator

Appendix A
International Alphabet 5 or ASCII Code

The ASCII (American Standard Code for Information Interchange) codes for numbers, alphabet letters and other common symbols

Decimal numbers	ASCII code in binary	in hex	Alphabetical characters	ASCII code in binary	in hex
0	0110000	30	@	1000000	40
1	0110001	31	A (a)	1000001	41 (61)
2	0110010	32	B (b)	1000010	42 (62)
3	0110011	33	C (c)	1000011	43 (63)
4	0110100	34	D (d)	1000100	44 (64)
5	0110101	35	E (e)	1000101	45 (65)
6	0110110	36	F (f)	1000110	46 (66)
7	0110111	37	G (g)	1000111	47 (67)
8	0111000	38	H (h)	1001000	48 (68)
9	0111001	39	I (i)	1001001	49 (69)
			J (j)	1001010	4A (6A)
Other			K (k)	1001011	4B (6B)
symbols			L (l)	1001100	4C (6C)
:	0111010	3A	M (m)	1001101	4D (6D)
;	0111011	3B	N (n)	1001110	4E (6E)
<	0111100	3C	O (o)	1001111	4F (6F)
=	0111101	3D	P (p)	1010000	50 (70)
>	0111110	3E	Q (q)	1010001	51 (71)
?	0111111	3F	R (r)	1010010	52 (72)
Space	0100000	20	S (s)	1010011	53 (73)
!	0100001	21	T (t)	1010100	54 (74)
"	0100010	22	U (u)	1010101	55 (75)
#	0100011	23	V (v)	1010110	56 (76)
$	0100100	24	W (w)	1010111	57 (77)
%	0100101	25	X (x)	1011000	58 (78)
&	0100110	26	Y (y)	1011001	59 (79)
'	0100111	27	Z (z)	1011010	5A (7A)
(	0101000	28			
)	0101001	29	[	1011011	5B
*	0101010	2A	\	1011100	5C
+	0101011	2B	]	1011101	5D
,	0101100	2C	↑	1011110	5E
-	0101101	2D	←	1011111	5F
.	0101110	2E			
/	0101111	2F			

Self-test Questions

1 Signals

1.1 List the factors that are considered in the selection of the bandwidth to be provided by an audio communication link.

1.2 Briefly discuss the reasons why a circuit provided for the transmission of music must have a wider bandwidth than a speech circuit.

1.3 Draw two curves to show how the thresholds of audibility and feeling vary for the average human ear. Explain the significance of the two points at which the two curves coincide with one another.

1.4 Figure Q1.1 shows the basic circuit of a bi-directional telephone link. Explain how it works. What is the main disadvantage of this simple circuit?

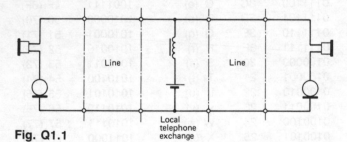

Fig. Q1.1

1.5 Which of the following factors are contributory to the bandwidth occupied by a television picture signal: (*a*) the number of lines per picture; (*b*) the number of frames per second; or (*c*) the duration and number of synchronization pulses per second?

1.6 A telegraphy signal consists of marks and spaces, each of which has a time duration of 100 ms. Calculate the baud speed of the signal.

1.7 Calculate the baud speeds of the two waveforms given in Fig. Q1.2.

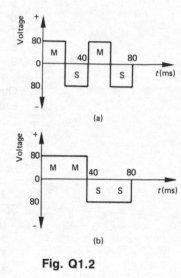

Fig. Q1.2

1.8 A teleprinter is operated at a speed of 75 bauds. Calculate the duration of a transmitted mark. Also calculate the maximum fundamental frequency of the waveform.

1.9 Explain why the full range of frequencies that the human voice is capable of generating and the human ear is able to hear is not transmitted over a communication system.

1.10 When music signals are transmitted over land lines the bandwidth provided is of the order of 50−10 000 Hz. Yet the bandwidth provided by medium-wave broadcast stations is only about 4500 Hz. Explain why this is so.

1.11 A data waveform has a bit rate of 600 bits/s. Calculate: (*a*) the baud speed; and (*b*) the duration of a bit.

1.12 The duration of a data bit is 0.833 ms. Calculate the bit rate.

1.13 Draw a data waveform that has a bit rate of 110 bits/s. Assume the ASCII code and the use of start and stop bits.

2 Frequency, Wavelength and Velocity

Short Exercises

2.1 Calculate the velocity of a signal of frequency 50 kHz and wavelength 5200 m.

2.2 A radio-frequency transmission line is λ/4 long at 300 MHz. Calculate the length of the line, in metres.

2.3 Calculate the wavelength of a 60 kHz signal propagating at the velocity of light.

2.4 A radio station has a maximum frequency tolerance of ± 10 parts in 10^6 and operates at 100 MHz. Calculate the maximum frequency change in hertz.

2.5 A radio station operating at a frequency of 90 MHz has a maximum allowable carrier frequency variation of ± 450 Hz. Calculate the frequency tolerance in parts per million.

2.6 Two radio stations operate at frequencies that are 15 kHz apart. Express the wavelength separation in metres. The lower frequency station operates at 1 MHz.

3 Modulation

Types of modulation

3.1 A 2 kHz sinusoidal waveform is pulse-amplitude modulated at a sampling rate of 20 kHz. Draw the modulated waveform.

3.2 Figure **Q3.1** shows several different waveforms. Identify each of them.

3.3 A train of 2 V amplitude, 10 μs-width rectangular pulses are pulse-amplitude-modulated by a 1-V peak sinusoidal signal. Draw the pam waveform.

3.4 The instantaneous voltage of a sinusoidal carrier wave is given by

$$v = V \sin(\omega t + \theta)$$

Which of these parameters is varied when the wave is: (*a*) frequency modulated; (*b*) amplitude modulated; and (*c*) phase modulated?

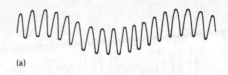
(a)

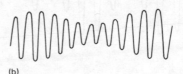
(b)

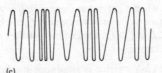
(c)

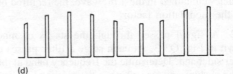
(d)

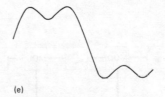
(e)

Fig. Q3.1

3.5 A 6 V carrier is modulated by a ramp waveform which changes linearly from 0 V to 3 V in 1 ms. Draw the resultant waveform when the carrier is: (*a*) amplitude modulated; and (*b*) frequency modulated.

3.6 What kind of waveform is shown in Fig. **Q3.2**? Determine (*a*) the carrier frequency; (*b*) the modulating frequency; and (*c*) the modulation factor.

3.7 A 72 kHz carrier wave is amplitude-modulated by a 2-kHz wave. Calculate the frequencies contained in the modulated waveform.

3.8 Determine the power in the sidefrequencies of an amplitude-modulated waveform if the total power is 10 kW and the depth of modulation is 60%

3.9 A 500-kHz carrier is amplitude-modulated by signals in the frequency band of 250−3000 Hz. Determine the frequencies contained in the a.m. wave and state the required bandwidth.

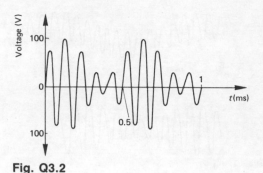

Fig. Q3.2

3.10 Show that when a carrier at frequency f_c is amplitude-modulated by a sine voltage at frequency f_m the bandwidth required is $B = 2f_m$ if $f_c > f_m$. Determine whether or not this relationship still applies if $f_m > f_c$.

3.11 A 50-kHz 12-V carrier is amplitude-modulated by a 6 V sine wave at a frequency of 2500 Hz. Calculate the frequencies contained in the a.m. wave, the required bandwidth and the modulation factor.

3.12 A signal passes through the stages of modulation shown in Fig. Q3.3. At each stage a filter passes only the lower sideband. Determine the frequency band at the output of the system and say whether the sideband is inverted or erect.

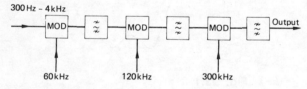

Fig. Q3.3

3.13 A 500 kHz carrier is amplitude-modulated by the band of frequencies 10–20 kHz. Determine the frequencies contained in (a) the lower sideband and (b) the upper sideband. What is the bandwidth necessary to transmit this signal?

3.14 A 28 V carrier wave is amplitude-modulated to a depth of 50%. Draw the waveform and calculate the maximum and minimum voltages of the wave.

3.15 The envelope of a sinusoidally modulated a.m. wave has a maximum value of 12 V and a minimum value of 3 V. Calculate the voltages of (a) the carrier, (b) the modulating signal, and (c) the upper sidefrequency.

3.16 The voltage of the lower sidefrequency of a sinusoidally modulated a.m. wave is 6 V. If the carrier voltage is 24 V, calculate (a) the modulating signal voltage and (b) the modulation factor.

3.17 A carrier wave has an unmodulated power of 100 W. Calculate the power in the lower sidefrequency if the carrier is sinusoidally modulated to a depth of 40%.

3.18 A 1000 W carrier wave is amplitude-modulated to a depth of 80%. What is the total power in the modulated wave?

3.19 Calculate the modulation factor of a carrier system if the voltage is 3 V and the signal voltage is 0.3 V.

3.20 Explain why the modulation index cannot be improved by amplifying the signal.

3.21 A 9 V carrier is amplitude-modulated by a sinusoidal signal with the result that its amplitude varies up to 12 V. Calculate the modulation factor and the amplitude of each sidefrequency.

3.22 If the wave described in 3.21 is applied across a 100 Ω resistor, calculate the power dissipated.

3.23 An 88-kHz carrier is amplitude-modulated by a 0–4 kHz signal and the lower sideband is selected. This sideband is then used to amplitude-modulate a 120 kHz carrier. Determine the frequencies contained in the lower sideband of the second modulation stage.

3.24 An a.m. waveform is applied across a 1000 Ω load. The powers produced by each of the sidefrequencies are 100 W and the carrier power is 2000 W. Calculate (a) the total power and (b) and the modulation factor.

3.25 Calculate the peak voltage at the output of a circuit when the carrier voltage is 0.01 V and the signal voltage is 1 mV.

3.26 Draw the frequency-spectrum graphs of both of the following: (a) dsbam and (b) ssbsc (upper sidefrequency). For both cases assume (i) carrier 12 V and 64 kHz, (ii) modulation signal 6 V and 0.3–3.4 kHz.

3.27 A 50-V carrier wave is amplitude-modulated by a sinusoidal voltage. Calculate the voltage of (a) the modulating signal, (b) the upper sidefrequency and (c) the lower sidefrequency. The modulation depth is 25%.

3.28 An fm wave has a rated system deviation of 10 kHz. What is the maximum frequency swing? What will be the frequency swing when the modulation voltage is one-half its maximum permitted value?

3.29 An fm system has a rated system deviation of 6 kHz and a maximum modulation frequency of 3 kHz. Calculate the necessary bandwidth to transmit the signal.

3.30 A number of fm channels are to be transmitted over a wideband cable having a bandwidth of 50 kHz. Each channel has an audio bandwidth of 0–4 kHz and a rated system deviation of 7 kHz. Calculate how many channels can be transmitted.

3.31 A 30-MHz carrier is frequency-modulated by a 15-kHz signal. The maximum frequency deviation of the carrier frequency is 70 kHz. Calculate (a) the modulation index and (b) the frequency swing of the modulated wave.

3.32 An fm system has a carrier frequency of 10 MHz. If the modulation frequency is 20 kHz and the available bandwidth is only 150 kHz, calculate (a) the deviation ratio and (b) the frequency swing.

3.33 An fm system has a rated system deviation of 6 kHz and a deviation ratio of 2. Calculate (a) the maximum modulation frequency and (b) the required bandwidth.

3.34 A 100-W carrier wave is frequency-modulated and its modulation index is then 5. Determine the power in the fm wave.

3.35 An fm system has a maximum frequency deviation of 20 kHz. If the modulating signal voltage is 50% of the maximum permitted value, calculate the frequency deviation.

3.36 Draw the fm wave that is produced by the modulating signal voltage waveform given in Fig. **Q**3.4.

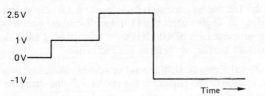

Fig. Q3.4

3.37 A 139-MHz carrier wave is frequency-modulated by a 3-kHz wave. List the component frequencies in the modulated wave if up to third-order components are present.

3.38 A 142-MHz carrier is frequency-modulated so that it has a modulation index of 6. If the modulation frequency is 10 kHz, calculate the frequency deviation of the carrier.

3.39 Explain the differences between the terms "modulation index" and "deviation ratio" and between "frequency deviation" and "rated system deviation".

3.40 Compare fm and am with respect to (a) the necessary bandwidth, (b) the power content and (c) the output signal-to-noise ratio.

4 Carrier Frequencies and Bandwidths

Table Q4.1

1	625-line TV broadcasts	**A**	405–525 kHz
2	Long-distance overseas sound broadcasting	**B**	25.6–26.1 MHz
		C	451–452 MHz
3	Long-distance point-to-point radio links	**D**	300–3400 Hz
		E	5.73–5.95 MHz
4	Ship-to-shore telegraphy	**F**	471.25–575.25 MHz
5	Mobile police radio		
6	MCVF telegraphy system over lines		

4.1 Base-mobile systems are operated by a number of concerns and organizations at frequencies in the vhf and uhf bands. List some of the more important of these. State typical channel bandwidths.

4.2 State the frequency bands used for satellite communications.

4.3 State the carrier frequencies and the bandwidths for each of the following: (a) sound broadcasting using frequency modulation, (b) the sound signal of 625-line TV, (c) the channels in a CCITT 12-channel group, and (d) a ship-to-shore long-distance radio-telephony link.

4.4 A taxi-cab firm uses radio to control its vehicles. State the frequency band in which the system might operate and give the probable channel bandwidths. What form of modulation would be used?

4.5 State the highest and the lowest carrier frequencies used by the CCITT 12-channel carrier group. If the lower sidebands are selected, calculate (a) the transmitted bandwidth and (b) the bandwidth per channel. Assume each channel occupies 0–1 kHz.

4.6 Match up the systems and the bandwidths quoted in Table Q4.1.

5 Basic Transmission Theory

5.1 A coaxial pair has a loss of 10 dB at a frequency of 1 MHz. Calculate its loss at 4 MHz, assuming that the dielectric losses are negligibly small.

5.2 A coaxial cable has a loss of 4 dB/km at 2 MHz. If dielectric losses are negligible, what is the loss of 2.5 km length of this cable at 4 MHz?

5.3 A generator of e.m.f. 50 V and internal impedance 600 Ω is applied to the sending-end terminals of a line having a characteristic impedance of 600 Ω and an attenuation coefficient of 1 dB/km. Calculate the current that flows in the correctly terminated load resistance if the line is 20 km long.

5.4 A correctly-terminated line has $Z_0 = 600$ Ω and $\alpha = 0.8$ dB/km. It is fed by a generator of e.m.f. 10 V and internal resistance 600 Ω. If the power dissipated in the load is 5 mW calculate the length of the line.

5.5 Explain the difference between the phase velocity and the group velocity of a line. What is meant by group delay and by group-delay/frequency distortion? Explain the effects that group-delay/frequency distortion may have upon (a) an analogue signal, and (b) a digital signal.

5.6 What is meant by the term characteristic impedance when it is applied to a transmission line? A line has $Z_0 = 750$ Ω and is correctly terminated. Determine its input impedance.

5.7 A signal consisting of a 5 kHz fundamental and its third harmonic are transmitted over a transmission line. The phase-change coefficient of the line is 10^0/km at 5 kHz and 24^0/km at 15 kHz. Calculate the group velocity of the signal.

5.8 A current of 12 mA flows into a 4 km length of transmission line. The current that flows in the correctly-terminated load is 4 mA. Calculate the attenuation coefficient of the line.

5.9 A cable has an attenuation coefficient of 1 dB/km at 1000 Hz and 1.8 dB/km at 3000 Hz. What will be the loss at each frequency of a 6 km length of this line?
The signal applied across the sending-end terminals of this line consists of a 10 V 1000 Hz fundamental plus 25% third harmonic. Calculate the voltage at each frequency across the correctly-terminated far end of the line.

5.10 A loss-free line has its far-end terminals connected to an impedance that is not equal to Z_0. Explain why standing waves will appear on the line. If the maximum and minimum voltages along the line are 20 V and 4 V respectively calculate the vswr.

6 Noise

6.1 An amplifier has a rectangular bandwidth of 100 kHz and a voltage gain of 150. Calculate the noise voltage at its output due to a 78 kΩ resistor connected across its input terminals. Will this be the total noise voltage at the output terminals? Assume the temperature to be 20°C and $k = 1.38 \times 10^{-23}$ J/K.

6.2 List and briefly explain the sources of noise in bipolar and field effect transistors.

6.3 Calculate the thermal noise voltage generated by a 500 kΩ resistor if the bandwidth is 2 MHz and the temperature is 17°C. $k = 1.38 \times 10^{-23}$ J/K.

6.4 Calculate the available noise power from a resistor in W/MHz if the temperature is 18°C. $k = 1.38 \times 10^{-23}$ J/K.

6.5 Explain why data circuits are more susceptible to noise than speech circuits. Give three sources of noise that may affect a data circuit but, unless severe, will not affect a speech circuit.

6.6 Calculate the signal-to-noise ratio at a point where the signal voltage is 10 V and the noise voltage is 100 mV.

6.7 The noise power level at a point where the signal-to-noise ratio is 37 dB is 100 μW. Calculate the signal power.

6.8 Calculate the noise power developed by a 56 kΩ resistor in a bandwidth of 100 kHz at a temperature of 17°C.

6.9 The noise level at the input to an amplifier is −40 dBm. If the input signal-to-noise ratio is 30 dB determine the amplifier gain necessary to give an output signal of +17 dBm.

6.10 The input noise to an amplifier is given by kTB watts. Calculate the input signal level needed to give an input signal-to-noise ratio of 45 dB. $T = 290$ K; $B = 2$ MHz.

7 Digital Signals and their Transmission over Lines

7.1 The arrival curve for a particular line has a silent interval of 10 μs, a transit time of 30 μs, and a steady current of 4 mA for 6 ms. Sketch the curve.

7.2 The Morse code signal dot, dash, dot is sent over a line using (a) single-current and (b) double-current operation. If the time duration of a dash is three times that of a dot sketch the sent current waveform for each case.

7.3 (a) Explain, with the aid of suitable sketches, how the waveform of the current at the end of a digital data line link can be constructed using the arrival curve.
(b) Use the arrival curve method to determine the waveform of the received current at the end of a line when the data waveform shown in Fig. Q7.1 is applied to the sending end of the line.

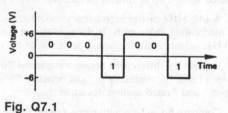

Fig. Q7.1

7.4 (a) Draw graphs of current and voltage against time at the sending and receiving ends of a digital data line when (i) single-current and (ii) double-current signals are used. (b) Use your graphs to explain why double-current working gives greater reliability and a faster speed of signalling.

7.5 List the relative merits of single-current and double-current working of a short digital data link.

7.6 Draw the digital data waveform which represents the binary number 011001.

7.7 A data signal is transmitted at 3200 bits/s. A character consists of four 1 bits followed by four 0 bits. Calculate the fundamental frequency of the waveform.

7.8 A 4800 bits/s signal is transmitted by a computer. What are the (a) minimum and (b) maximum fundamental frequencies of the waveform? What other frequencies may also be present?

7.9 Draw a sine wave of frequency 300 Hz and peak value 6 V. Then draw the third and fifth harmonics of this wave with peak values of 2 V and 1 V respectively. Sum the three waves to obtain the resultant waveform.

7.10 The duration of each bit on a line is 1.042×10^{-4} s. Calculate the bit rate.

7.11 Explain why long-distance signalling employs voice-frequency signals instead of d.c. even though the latter are generated by the telephone dial.

7.12 Explain how group-delay/frequency distortion has an effect on (a) analogue and (b) digital signals.

7.13 List, and briefly explain, four items in an analogue telephone network that will not transmit the d.c. component of a data waveform.

7.14 (a) The digital data waveform shown in Fig. Q7.2 has its d.c. component removed. Sketch the resultant waveform. (b) Why would the d.c. component of a digital data waveform be removed if the signal were transmitted over the pstn?

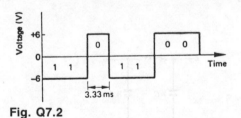

Fig. Q7.2

7.15 (a) Explain, with the aid of waveform diagrams, the effects of attenuation and group-delay/frequency distortion on a digital data waveform.
(b) Show, with the aid of block diagrams, how the effects of (a) can be overcome using (i) modems and amplifiers, (ii) regenerators.

8 Digital Modulation

8.1 (a) Outline a method by which data may be transmitted over an audio-frequency circuit at a bit rate of 600 bits/s.
(b) How can errors in the received data be detected?

8.2 Explain, with the aid of waveform sketches, what is meant by frequency shift modulation. A 1500 Hz carrier wave is frequency shift modulated by a 600 bits/s data waveform.
(a) What is the maximum fundamental frequency of the data waveform?
(b) What bandwidth is required for the transmission of the frequency-shift waveform if (i) the fundamental component only is transmitted, (ii) the third harmonic is also transmitted?
(c) Why is it possible to transmit only the first-order side-frequencies and what is the advantage gained by so doing?

8.3 What is meant by the term *dibit* used in data communications? How can dibits be represented by (a) changes in the phase angle of a carrier wave, (b) changes in the amplitude of a carrier wave? State the advantage which is gained by the use of multi-phase or multi-level transmission.

8.4 Draw the waveform of a carrier wave that has been frequency-shift modulated by the data signal 1011001 at a bit rate of 1200 bits/s. Label the diagram with the frequencies that are transmitted and state the nominal carrier frequency. Which data waveform will have the highest fundamental frequency and what is then the occupied bandwidth?

8.5 Explain how the performance of a frequency-shift system can be adversely affected by line attenuation and group-delay/frequency distortion. Why is a 1200 bits/s system more likely to be affected than a 600 bits/s system?

8.6 Draw the block diagram of a modem, showing both transmitting and receiving sections, and briefly explain its operation. Discuss the factors that influence the choice of carrier frequency.

8.7 What is the bandwidth of a commercial-quality speech circuit? Why cannot all of this bandwidth be employed for data transmission? State what bandwidth is available (a) for a frequency-shift system, (b) for a differential phase system. Why is dsb amplitude modulation not generally used for data transmission?

8.8. How does frequency shift modulation differ from frequency modulation? Illustrate your answer with waveform sketches. Calculate the deviation ratio for a 600 bits/s frequency-shift data system. What bandwidth is needed for such a system?

8.9 What is meant by *vestigial sideband amplitude modulation* and why is it used in some data communication systems? Give brief details of the 48 kilobits/s vsb system used in the United Kingdom and draw the spectrum diagram of this system.

9 Attenuators, Equalizers and Filters

9.1 Draw the circuit diagram of a basic low-pass filter and draws its (a) ideal and (b) practical attenuation−frequency characteristics. What is the general name given to this kind of filter?

9.2 A signal occupying the frequency band 50 Hz−25 kHz is applied to the input of the two cascaded filters shown in Fig. Q9.1. Determine the bandwidth of the output signal.

9.3 If the cut-off frequencies of the two filters shown in Fig. Q9.1 are switched over, what will then be the output bandwidth?

Fig. Q9.1

9.4 The attentuation-frequency characteristic of a filter is given in Table Q9.1. Plot the characteristic and state what kind of filter it is.

Table Q9.1

Frequency (Hz)	10	30	100	300	1000	3000	10 000	30 000
Attenuation(dB)	30	20	10	1	1	2	20	30

9.5 (a) What kind of filter is shown in Fig. Q9.2? (b) What is meant by the term "constant-k"? (c) Briefly explain how the filter works. (d) Sketch a typical loss—frequency characteristic.

9.6 Figure Q9.3 shows the attenuation—frequency characteristic of a filter, (a) What kind of filter is it? (b) What does f_c stand for? (c) Sketch the ideal characteristic of this filter.

9.7 The low-pass filter shown in Fig. Q9.4 is to have its loss just outside the passband increase more rapidly. Indicate how this can be done and say what the technique is called.

9.8 Determine the band of frequencies that appears at the output of Fig. Q9.5.

9.9 List the relative merits of LC and crystal filters.

9.10 Explain why there is a need for filters in communication networks.

9.11 A T attenuator is to have 9 dB loss and a characteristic resistance of 140 Ω. Design the circuit.

9.12 A π attenuator is to have 9 dB loss and a characteristic resistance of 140 Ω. Design the circuit.

9.13 The input signal to a line amplifier is at a level of -7 dBm. The amplifier has a gain of 27 dB and its output level is to be $+10$ dBm. Determine the loss of an attenuator to be placed at the input to the amplifier to achieve the requirements.

9.14 Design a T attenuator to have 15 dB loss and a characteristic resistance of 75 Ω.

9.15 Design a π attenuator to have 20 dB loss and a characteristic resistance of 600 Ω.

9.16 Draw the circuit of (a) an audio attenuation equalizer, and (b) a low-pass filter. Explain how they differ from one another.

9.17 The group-delay frequency characteristic of a circuit is described by the data given in Table Q9.2. Write down the table that would describe the characteristics of the ideal group-delay equalizer for this circuit.

Table Q9.2

Frequency (Hz)	400	800	2000	3000
Delay (ms)	0.01	0.005	0.07	0.30

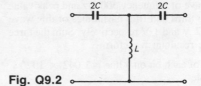

Fig. Q9.2

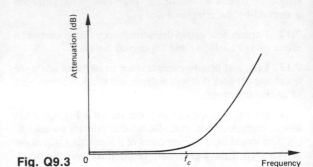

Fig. Q9.3

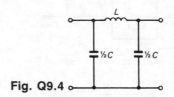

Fig. Q9.4

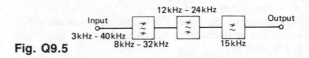

Fig. Q9.5

9.18 Draw a typical attenuation—frequency characteristic for an audio-frequency line of 6 km length. Hence explain the need for attenuation equalization.

10 Two-wire and Four-wire Circuits

10.1 Draw the block diagram of a four-wire circuit and briefly explain its operation.

10.2 Discuss the reasons why a four-wire circuit may *howl* or *sing*.

10.3 What is meant by *noise* in a transmission system? Give four sources of noise.

10.4 An audio circuit is to be amplified using a single amplifier. Explain why this amplifier should not be installed at either end of the circuit but approximately in the middle.

10.5 What are the components in an amplified four-wire circuit that prevent the passage of a digital data signal?

10.6 The lines used to connect the amplifiers used in the circuit of Fig. Q10.1 have a loss of 1 dB/km at 500 Hz and 1.2 dB/km at 1500 Hz. Calculate the output signal power in dBm at each frequency when the input signal is 0.1 mW at 500 Hz and 0.05 mW at 1500 Hz.

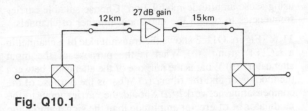

Fig. Q10.1

10.7 The point marked C in Fig. Q10.2 is short-circuited. Determine whether or not the circuit will become unstable.

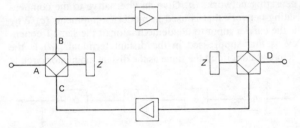

Fig. Q10.2

10.8 A four-wire circuit is set up over an unloaded cable to have an overall loss of 3 dB. Owing to a fault, at some later date, one section of this cable has to be replaced by a length of loaded cable. Explain whether the circuit will become potentially unstable.

10.9 Draw the circuit of a four-wire link having (a) end repeater stations A and D, (b) intermediate repeater stations B and C, (c) cable losses A−B = 25 dB, B−C = 20 dB and C−D = 18 dB, (d) two-wire input level at A = −3 dBm and overall loss = 3 dB. Use attenuators at amplifier inputs and outputs where necessary.

10.10 State a typical signal-to-noise ratio for an amplified audio speech circuit.

10.11 Why would a digital data signal not transmit over the circuit shown in Fig. Q10.2?

10.12 Draw the circuit of a two-wire/four-wire terminating unit and explain its operation.

10.13 State the differences between the balance impedances required for (a) two-wire and (b) four-wire circuits.

10.14 Explain clearly why the audio signal travelling along an audio circuit must be amplified at regular intervals. Why cannot a pulse regenerator be employed instead?

10.15 Why is two-wire operation of an amplified circuit only possible for the shorter-length circuits?

10.16 Explain the function of a group switching centre in a trunk telephone network.

10.17 Draw the layout of a typical telephone exchange area. Make clear the reasons why pillars and cabinets are used.

10.18 Show, with the aid of a block diagram, how a conventional amplifier which only works in one direction can be used in a bi-directional two-wire circuit.

10.19 In a two-wire amplifier using conventional amplifiers and two hybrid coils the amplifier gains are each 15 dB. If the input signal level to the amplifier is −4 dBm, what is the output signal level?

10.20 The amplifiers shown in Fig. Q10.3 each have a gain of 27 dB. The amplifiers are to have an output level of +10 dBm when the level of the applied signal at the two-wire terminals is −3 dBm. If the overall loss of the circuit, in each direction, is to be 3 dB, calculate the required attenuation of each of the six attenuators.

10.21 For the circuit given in Fig. 10.3 indicate the points at which an attenuation equalizer might be fitted.

10.22 List five sources of noise that are likely to affect a line system. Suggest how the effect of each may be reduced.

10.23 What is meant by *crosstalk*? Where might it occur? How can it be reduced?

10.24 Draw, and explain, an international radio telephony circuit that uses a carrier in the high-frequency band. State the bandwidth provided.

10.25 List the relative merits of two-wire and four-wire operation of an amplified circuit.

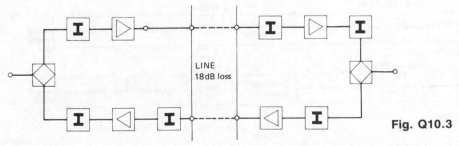

LINE
18 dB loss

Fig. Q10.3

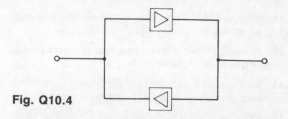

Fig. Q10.4

10.26 Explain why the circuit given in Fig. Q10.4 will not operate correctly. How must the circuit be modified to make it work?

11 Frequency-division Multiplex Systems

11.1 Figure Q11.1 shows partially the spectrum diagram for a CCITT 12-channel group. Insert the carrier frequencies used per channel and also the upper and lower frequencies transmitted per channel.

11.2 Figure Q11.2 shows the channel-translating equipment of one channel in a CCITT 12-channel group. State the function of each block.

11.3 Draw the circuit of a Cowan modulator and explain its operation.

11.4 Why are double-balanced modulators sometimes used instead of single-balanced modulators? State one situation where a double-balanced modulator would be used.

11.5 A line has a bandwidth of 0–20 kHz. A number of 4 kHz bandwidth channels are to be transmitted over the line using ssbsc amplitude-modulation. Choose suitable carrier frequencies to obtain the maximum number of channels.

11.6 Figure Q11.3 shows the transmit side of a channel in a CCITT group. (*a*) What is the purpose of the input attenuator? (*b*) What is the purpose of the attenuator between the modulator and the filter? (*c*) What is the purpose of the compensating network? (*d*) Should the carrier input to the modulator be of greater amplitude than the signal amplitude and, if so, by how much?

11.7 Figure Q11.4 shows the receive side of a channel in a CCITT 12-channel group. (*a*) State the function of the compensating network. (*b*) Give an alternative to the compensating network that is used in modern equipment. (*c*) Why is the carrier input to the demodulator of the same frequency as that suppressed in the distant terminal? (*d*) Is the demodulator circuit the same as the distant modulator circuit?

Fig. Q11.1

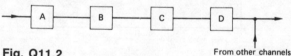

Fig. Q11.2

From other channels

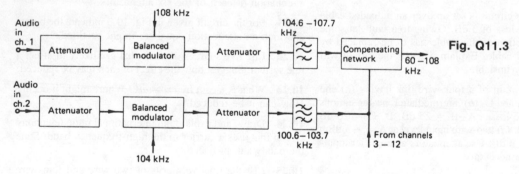

Fig. Q11.3

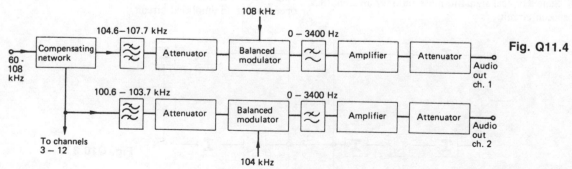

Fig. Q11.4

11.8 The bandwidth per channel in a CCITT 12-channel group is 300–3400 Hz, yet the carrier frequencies are spaced 4000 Hz apart. Why is there an apparent waste of 900 Hz?

11.9 Explain, with the aid of a block diagram, how a 900-channel hypergroup can be formed.

12 Pulse-code Modulation and Time-division Multiplex

12.1 Draw, and explain, the block schematic diagram of a three-channel pcm system. Ensure that the common line shows signals from each of the three different channels being transmitted.

12.2 The signal shown in Fig. Q12.1 is sampled at the times, *a, b, c, d* and *e*. Show the binary-coded representation of each sample.

12.3 The signals shown in Figs Q12.2(a) and (b) are applied respectively to the channels of a two-channel pcm system. Sketch the waveform on the common line. Assume eight sampling levels.

12.4 Explain why accurate synchronization of a tdm system is essential.

12.5 Draw the frame structure of a 30-channel pcm system. State which of the channels are available for the transmission of speech. What do the other channels carry?

12.6 Explain why the first step in the production of a pcm signal is the production of a pam waveform.

12.7 Explain how quantization noise is generated within a pcm system. How can this form of noise be reduced and what is the disadvantage of so doing?

12.8 Explain why the range of signal amplitudes that a pcm system can handle is divided into a number of sampling levels. Why is the number of levels provided always some multiple of 2?

12.9 Successive samplings of a signal produce quantized levels of 3, 37, 53, 113, 60 and 28. Draw the binary waveform that is transmitted.

12.10 A pcm system has 1024 sampling levels. Calculate the number of pulses transmitted per second if the sampling frequency is 15 kHz.

12.11 The 30-channel pcm system uses pulses that are 0.488 μs wide. If the sampling frequency is 8 kHz and there are 256 sampling levels, calculate how many channels can be transmitted. Why are only 30 of these channels employed for speech transmission?

12.12 Calculate the line bit rate of a 30-channel pcm system.

12.13 Refer to Fig. Q12.3. (*a*) What is a muldex? (*b*) Quote the higher-order bit rates that are recommended by the CCITT. (*c*) Show how Fig. Q12.3 can be made the basis of each of the higher-order systems in (*b*).

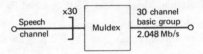

Fig. Q12.3

12.14 List the relative merits of fdm and tdm.

12.15 A 1500-Hz sine wave of peak value 10 V is transmitted over a pcm system. The sampling frequency is three times the minimum allowable. Determine (*a*) the instantaneous values of the sine wave at each sampling instant, assuming the first sample is taken at time $t = 0$. If the system caters for a maximum input voltage of ± 12 V and uses 128 sampling levels, write down the binary train transmitted to line to represent each sample.

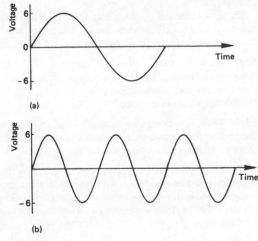

Fig. Q12.2

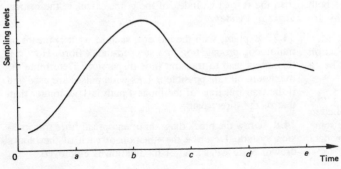

Fig. Q12.1

12.16 Which quantum level represents a sampled amplitude of 0.54 V if each equally spaced quantum level corresponds to a voltage step of 0.01 V?

12.17 The number of quantum levels employed in a pcm system is increased from 128 to 512. Calculate the percentage reduction in the quantization error that results.

12.18 The basic block diagram of a pcm system consists of the following blocks: (*a*) sampling gates; (*b*) encoder; (*c*) transmission line; (*d*) decoder; (*e*) demultiplexor; and (*f*) channel low-pass filters. Sketch the block diagram and indicate the waveform at the output of each block.

13 Data Circuits

13.1 Briefly explain the features of the British Telecom services Kilostream, Megastream, Satstream and Packetstream.

13.2 Explain the method employed to transmit data over (*a*) a 64 kilobits/s circuit, and (*b*) a 2 Megabits/s circuit.

13. 3 What is the difference between a concentrator and a multiplexer? Draw the block diagram of a data system in which two leased circuits and two links routed via the pstn are connected to a multiplexer for transmission over a high-speed data link to another multiplexer which is connected to the front-end processor of a computer. Label each data link drawn with a typical bit rate.

13.4 List the factors which determine whether a leased point-to-point circuit or the pstn should be used to connect a data terminal to a computer centre. Discuss how the decision will be affected by the use of (*a*) multiplexers, (*b*) concentrators.

13.5 Explain the need for point-to-point data links (*a*) between a computer and a data terminal and (*b*) between two computers.

Show, with the aid of block diagrams, how two computers can be linked together when the distance between them is (*a*) very short, (*b*) a few kilometres, (*c*) 40 km.

13.6 Draw the block schematic diagram of a centralized computer network. Include examples of the use of both multiplexers and concentrators in your diagram. Why are branching points often used?

13.7 What is a computer bureau? Explain the effects on the traffic flow through a telephone exchange that has a number of lines to a computer bureau.

13.8 Draw sketches to explain what is meant by the terms two-wire operation and four-wire operation when applied to (*a*) amplified audio-frequency lines (*b*) data links. Give some reasons why a four-wire circuit is often used for data transmission.

13.9 Explain the need for point-to-point links in a data network. What are the advantages of leasing a private circuit? Give some possible ways in which a point-to-point link may be routed. Explain why modems and amplifiers are used for most long distance data links.

13.10 Twelve 110 bits/s data channels are simultaneously transmitted over a line using time division multiplex. What is the bit rate on the line?

13.11 Six 110 bits/s data links are connected to a concentrator which has three output links. What is the bit rate on the output links?

13.12 State the advantages of voice-frequency operation of a data link compared with d.c. signalling over long distances.

13.13 Discuss briefly the circumstances which may decide that a data link should be rented full-time and not dialled-up as required.

14 Optical Fibre Systems

14.1 A ray of light travels through a medium of refractive index 1.68. Calculate the minimum angle of refraction that will give total internal refraction.

14.2 Draw, and briefly explain, the block diagram of a pulse regenerator for use in an optical fibre system.

14.3 State the range of wavelengths that are employed for optical fibre communications. List two light sources and two light detectors for these wavelengths.

14.4 List the advantages of optical fibre systems over coaxial cable systems.

14.5 The repeaters in a conventional coaxial cable system need to be spaced at approximately 2 km intervals along a route. The repeaters in an optical fibre system are spaced at about 10 km intervals. List, and explain, the advantages arising from this greater repeater spacing.

14.6 Describe how light energy can be propagated along an optical fibre by means of reflection and refraction. If, for a particular fibre, the cladding has a refractive index 1% less than the refractive index of the core, calculate the critical angle of incidence.

14.7 Explain, with the aid of a sketch, what is meant by multimode propagation in a stepped-index fibre. How can this effect lead to transmit time dispersion? Determine the maximum bit rate possible if 1 μs wide pulses are used and if the transmit time of the longest path is 1 μs longer than that of the direct path.

14.8 Draw the block diagram of an optical fibre transmission system. Describe the function of each of the blocks drawn. State what kind of modulation is employed.

14.9 List the sources of loss in an optical fibre system. Briefly explain each of them. Draw a typical attenuation/wavelength characteristic and say why particular wavelengths only are so far used for optical fibre systems.

14.10 The operating wavelength of an optical fibre system is 0.85 μm. Determine the dimensions of the cable cores shown in Figs. 14.7(a) and 14.8(a) in terms of wavelengths.

14.11 An optical fibre cable has a bandwidth—distance product of 2 GHz km. Determine (a) the maximum frequency that can be used over a distance of 20 km, (b) the maximum cable length possible for transmission at 500 MHz.

14.12 Explain the terms stepped-index and graded-index when applied to an optical fibre.

Self-test Answers

1.5 All three **1.6** 10 **1.7** 50, 50 **1.8** 13.3 ms, 37.5 Hz **1.11** 600, 1.67 ms **1.12** 1200 bits/s

2.1 2.6×10^8 m/s **2.2** 0.25 m **2.3** 5 km **2.4** ± 1 kHz **2.5** ± 5 p.p.m. **2.6** 4.43 m

3.2(*a*) Beat, (*b*) amplitude modulation, (*c*) frequency modulation, (*d*) pulse-amplitude modulation, (*e*) fundamental plus third harmonic **3.4**(*a*) ω, (*b*) V, (*c*) θ **3.6** dsbam, 10 kHz, 2 kHz, 50% **3.7** 70 kHz, 72 kHz, 74 kHz **3.8** 1.8 kW **3.9** 497−499.75 kHz, 500 kHz, 500.25−503 kHz **3.11** 47.5 kHz, 50 kHz, 52.5 kHx, 0.5 **3.12** 236−239.7 kHz, inverted **3.13** 480−490 kHz, 510−520 kHz, 40 kHz **3.14** 42 V, 14 V **3.15** 7.5 V, 4.5 V, 2.25 V **3.16** 12 V, 0.5 **3.17** 4 W **3.18** 1320 W **3.19** 10% **3.20** 0.33, 2 V **3.22** 0.45 W **3.23** 32−36 kHz **3.24** 2200 W, 0.447 **3.25** 11 mV **3.27** 12.5 V, 6.25 V **3.28** 20 kHz, 10 kHz **3.29** 18 kHz **3.30** 2 **3.31** 4.67, 140 kHz **3.32** 2.75, 110 kHz **3.33** 3 kHz, 18 kHz **3.34** 100 W **3.35** 10 kHz **3.37** 139 MHz, 139 MHz± 3 kHz, 139 MHz± 6 kHz, 139 MHz± 9 kHz **3.38** 60 kHz

4.5 60−108 kHz, 4 kHz

5.1 20 dB **5.2** 14.14 dB **5.3** 4.167 mA **5.4** 3.75 km **5.6** 750 Ω **5.7** 257.14 km/s **5.8** 2.39 dB/km **5.9** 5 V, 0.72 V **5.10** 5

6.1 1.68 mV **6.3** 126.5 μV **6.4** 4×10^{-15} W/MHz **6.6** 40 dB **6.7** 500 mW **6.8** 4×10^{-16} W **6.9** 27 dB **6.10** 0.25 nW

7.7 400 Hz **7.8** 0 Hz, 2400 Hz **7.10** 9600 bits/s

8.8 0.67, 600 Hz

9.2 zero **9.3** 20−25 kHz **9.8** 12−15 kHz **9.13** 10 dB

10.6 −18 dBm, −26.4 dBm **10.7** No **10.10** 30 dB **10.19** +3 dBm **10.20** 10 dB, 9 dB, 12 dB for each direction

12.10 150×10^3 s^{-1} **12.11** 32 **12.12** 2048 kbits/s **12:16** 54 **12.17** 75%

13.10 1320 bits/s **3.11** 110 bits/s

14.1 36^0 32^1 **14.7** 250 kbits/s **4.11** 100 kHz, 4 km

Index

BANFF AND BUCHAN COLLEGE OF FURTHER EDUCATION

BANFF AND BUCHAN COLLEGE OF FURTHER EDUCATION

BANFF AND BUCHAN COLLEGE OF FURTHER EDUCATION

BANFF AND BUCHAN COLLEGE OF FURTHER EDUCATION

BANFF AND BUCHAN COLLEGE OF FURTHER EDUCATION